Coastal and Oceanic Landforms, Development and Management

Coastal Zones

Management, Assessment and Current Challenges

COASTAL AND OCEANIC LANDFORMS, DEVELOPMENT AND MANAGEMENT

NOTICE TO THE READER

The Publisher has taken reasonable care in the preparation of this book, but makes no expressed or implied warranty of any kind and assumes no responsibility for any errors or omissions. No liability is assumed for incidental or consequential damages in connection with or arising out of information contained in this book. The Publisher shall not be liable for any special, consequential, or exemplary damages resulting, in whole or in part, from the readers' use of, or reliance upon, this material. Any parts of this book based on government reports are so indicated and copyright is claimed for those parts to the extent applicable to compilations of such works.

Independent verification should be sought for any data, advice or recommendations contained in this book. In addition, no responsibility is assumed by the publisher for any injury and/or damage to persons or property arising from any methods, products, instructions, ideas or otherwise contained in this publication.

This publication is designed to provide accurate and authoritative information with regard to the subject matter covered herein. It is sold with the clear understanding that the Publisher is not engaged in rendering legal or any other professional services. If legal or any other expert assistance is required, the services of a competent person should be sought. FROM A DECLARATION OF PARTICIPANTS JOINTLY ADOPTED BY A COMMITTEE OF THE AMERICAN BAR ASSOCIATION AND A COMMITTEE OF PUBLISHERS.

Additional color graphics may be available in the e-book version of this book.

Library of Congress Cataloging-in-Publication Data

ISBN: 978-1-63485-611-9

Published by Nova Science Publishers, Inc. † New York

COASTAL AND OCEANIC LANDFORMS, DEVELOPMENT AND MANAGEMENT

COASTAL ZONES

MANAGEMENT, ASSESSMENT AND CURRENT CHALLENGES

GINA MORAN
EDITOR

New York

CONTENTS

PREFACE

This book provides up-to-date research on the management, assessment and current challenges of coastal zones. Chapter One reviews the Coastal Area Management Programme (CAMP) of the Mediterranean Action Plan (MAP). Chapter Two introduces atypical example accounting for a comparative analysis between wetlands in Lebanon with different hydrology and aquifer characteristics. Chapter Three aims to account for procedures taken by coastal, insular, and archipelagic states when claiming the extension of its oceanic limits.

Chapter 1 – The Coastal Area Management Programme (CAMP) of the Mediterranean Action Plan (MAP) is a "country-driven" programme aimed at helping the Mediterranean Countries in their efforts towards sustainable development. Within it, the CAMP Italy Project is part of the Italian broader strategy for protection, management and sustainable development of the marine and coastal zones within the Barcelona Convention for the protection of the marine environment and the Mediterranean coastal region. Moreover, being a multi area project, it represents a unique experience within the CAMPs scenario. The Agreement for the CAMP Italy Project was signed in March 2014 by the Italian Ministry of Environment, Land and Sea (IMELS), UNEP-MAP, and three Italian Regions (Emilia-Romagna, Sardinia and Tuscany). The major goal of the CAMP Italy Project is to encourage the implementation of national strategies for Integrated Coastal Zone Management (ICZM) in Italy, based on pilot actions in several coastal areas of the sub mentioned Italian Regions. The Feasibility Study of the Project revealed a complex scenario that hinders the integrated management of coastal zones, mainly because of the fragmentation existing among the different levels of government. In particular, the main problems in the CAMP Areas are related to the absence of a centralized responsible authority for the coastal management. In fact, although acknowledging that IMELS has a focal role in the field of integrated coastal zone management, nowadays Italy has neither a coastal Act, nor a dedicated coastal governance body or management Agency. This is also linked to the fact that Italy has signed but not ratified the ICZM Protocol of the Barcelona Convention. This means that to date the Protocol and its principles, which are at the core of the CAMP Project, are not part of the national law, although it has been adopted by the EU and thus part of the EU *acquis communautaire*. The CAMP Italy includes actions aimed at reducing critical processes (coastal erosion, loss of biodiversity, pollution, etc.), managing natural resources in a sustainable way, and conserving natural habitats and biodiversity. CAMP Italy is structured in 14 Individual Activities (pilot projects) and 4 Horizontal Activities aimed at four goals: Coordination and dissemination of the results; Capacity building; Data collection and management; Institutional coordination and public participation. The chapter describes the Project and the adopted solutions to apply and achieve ICZM at both regional and Italian national level.

Chapter 2 – The unique landscape, where wetlands are located, needs to be monitored and reserved, since their continuity depends on the balanced input-output water system. The hydrologic setting of wetlands is always creating unique ecosystems for flora (e.g., pollens, etc.) and fauna (e.g., migratory birds, etc.). In Lebanon, wetlands are well pronounced, and they own different dimensions ranging from several hundreds of square meters up to few square kilometers. Lately, the Lebanon's wetlands have been subjected to physical and anthropogenic stresses, notably the water feeding becomes crucial, and this makes some wetlands losing their hydrologic criteria (i.e., water-saturated terrain). The changing climatic conditions, with a special emphasis to the changing rainfall patterns and temperature increase, have a primary role in affecting wetlands in Lebanon, notably those with fragile hydrologic balance. This has been also reflected on the major terrain components (i.e., land cover/use) which have been also changed over the last few decades. Hence, the area of these wet landforms is reduced due to the decreased volume of feeding water, and this has been exacerbated in the view of uncontrolled human interference. In this study, two wetlands along the Lebanese coastal zone were investigated using different tools and approaches for analysis. Even though, both selected wetlands are extending along the coast, yet they have different hydrologic regime, and thus they have been affected at different levels. This study introduces atypical example accounting for a comparative analysis between wetlands in Lebanon with different hydrology and aquifer characteristics.

Chapter 3 – The diversity of uses of the oceanic space and the old conception of "freedom of the seas doctrine" has compelled an adaptation of the political-economic and spatial legal framework for the oceanic ecosystem. This changing of direction encompassed the international system and has given rise to a new geopolitics for the legal framework of oceanic spaces around the globe, in terms of the United Nations Convention on the Law of the Sea (UNCLOS). Every coastal state who has signed the Convention acquired the right to claim its underwater territory or extended continental shelf up to 350 nautical miles from its coast. This study aims to account for procedures taken by coastal, insular, and archipelagic states when claiming the extension of its oceanic limits.

In: Coastal Zones
Editor: Gina Moran

ISBN: 978-1-63485-611-9

Chapter 1

TOWARDS THE IMPLEMENTATION OF INTEGRATED COASTAL ZONE MANAGEMENT IN THE MEDITERRANEAN: THE CAMP ITALY PROJECT

Alessio Satta[1,*], Beatrice M. S. Giambastiani[1], Daniela Addis[1-2], Ilda Mannino[1], Irene Sanguineti[1] and Elisa Ulazzi[1]

[1]CoNISMa – National Inter-University Consortium for Marine Sciences, Rome, Italy

[2]Priority Actions Programme/Regional Activity Centre, Croatia

ABSTRACT

The Coastal Area Management Programme (CAMP) of the Mediterranean Action Plan (MAP) is a "country-driven" programme aimed at helping the Mediterranean Countries in their efforts towards sustainable development. Within it, the CAMP Italy Project is part of the Italian broader strategy for protection, management and sustainable development of the marine and coastal zones within the Barcelona Convention for the protection of the marine environment and the Mediterranean coastal region. Moreover, being a multi area project, it represents a unique experience within the CAMPs scenario. The

[*] Corresponding author: Email: satta.alessio@gmail.com.

Agreement for the CAMP Italy Project was signed in March 2014 by the Italian Ministry of Environment, Land and Sea (IMELS), UNEP-MAP, and three Italian Regions (Emilia-Romagna, Sardinia and Tuscany). The major goal of the CAMP Italy Project is to encourage the implementation of national strategies for Integrated Coastal Zone Management (ICZM) in Italy, based on pilot actions in several coastal areas of the sub mentioned Italian Regions. The Feasibility Study of the Project revealed a complex scenario that hinders the integrated management of coastal zones, mainly because of the fragmentation existing among the different levels of government. In particular, the main problems in the CAMP Areas are related to the absence of a centralized responsible authority for the coastal management. In fact, although acknowledging that IMELS has a focal role in the field of integrated coastal zone management, nowadays Italy has neither a coastal Act, nor a dedicated coastal governance body or management Agency. This is also linked to the fact that Italy has signed but not ratified the ICZM Protocol of the Barcelona Convention. This means that to date the Protocol and its principles, which are at the core of the CAMP Project, are not part of the national law, although it has been adopted by the EU and thus part of the EU *acquis communautaire*. The CAMP Italy includes actions aimed at reducing critical processes (coastal erosion, loss of biodiversity, pollution, etc.), managing natural resources in a sustainable way, and conserving natural habitats and biodiversity. CAMP Italy is structured in 14 Individual Activities (pilot projects) and 4 Horizontal Activities aimed at four goals: Coordination and dissemination of the results; Capacity building; Data collection and management; Institutional coordination and public participation. The chapter describes the Project and the adopted solutions to apply and achieve ICZM at both regional and Italian national level.

Keywords: integrated coastal zone management, CAMP, governance, institutional coordination, participation, communication, data management, coastal planning

1. Introduction

The Coastal Area Management Programme (CAMP) of the Mediterranean Action Plan (MAP) promotes pilot projects in coastal areas of the Signatory Countries of the Barcelona Convention (Council Decision 2010/631/EU), in which Integrated Coastal Zone Management (ICZM) is the main tool for achieving sustainable development. Thereby, since 1989, when the Programme was approved by Contracting Parties (CPs), CAMP is a "country-driven" Programme aimed at helping the Mediterranean Countries in their efforts towards sustainable development. The signing of the ICZM Protocol (Official Journal L034, 2009), within the Barcelona Convention Framework (in Madrid on 21 January 2008) represents a very significant milestone in this process. The activities of the Programme are concentrated in the coastal areas of Mediterranean Countries. The interested Country is proposing its own project, and when it is being approved by the Contracting Parties at their biannual meetings, the preparatory activities can start. The projects are being developed by the local and national experts with the assistance of UNEP MAP and its Components, each one of them being involved in the activities falling within its area of expertise. The Priority Actions Programme Regional Activity Centre (PAP/RAC) is co-ordinating MAP efforts in each CAMP project.

The main goals behind the implementation of the CAMP projects in Mediterranean coastal regions are the following:

- Develop strategies and procedures at local, regional and national levels for sustainable development, environmental protection and the rational use of coastal and marine areas, which may be useful as experiences that form the basis of the formulation of Mediterranean strategies for sustainable development;
- Identify, adapt and test Integrated Coastal Zone Management methodologies and tools in real geographical contexts;
- Contribute to build capacities building of public institution personnel at the national, regional and local levels;
- Promote a more extensive use at the national, regional and local levels of the experience reached through the CAMP Programme and its individual activities, and create the conditions for the capitalization of these activities.

1.1. CAMP Italy Project

The proposal to start up a Coastal Area Management Programme (CAMP) Project on the Italian coastal zone was presented by the Italian authorities to UNEP-MAP in July 2007 and was taken into consideration and approved during the 15th meeting of the Contracting Parties to the Barcelona Convention (Almeria, Spain, 15-18 January 2008). The Agreement for the CAMP Italy Project was signed on 27 March 2014 on the basis of the preparatory activities that took place prior to the signature between the Italian Ministry of

Environment, Land and Sea (IMELS) and UNEP-MAP. A 2-year implementation project was envisaged (May 2014 – May 2016) as well as an additional implementation of post project activities in the following two years. The final project area includes several coastal strips and involves three Italian Regions: Emilia-Romagna, Sardinia and Tuscany. Existing spatial planning instruments were carefully considered for the three Italian Regions in order to select the most suitable areas, coastal types, habitat types and distribution, geographical features, administrative organization, degree of human settlement and economic activities. Within the three Regions five areas (CAMP areas) were seleceted (Figure 1) based on the following concepts:

Figure 1. The five CAMP Italy Project Areas.

- Presence of territorial planning instruments and/or experience on management of coastal areas to ensure the feasibility of the CAMP Project with existing resources;
- Sites with significant naturalistic and environmental value to preserve their biodiversity (i.e., presence of natural protected areas);
- Areas with specific geographical and physical characteristics, as well as anthropogenic pressures, which can be at risk from current and future effects of climate change, such as flooding by sea-level rise, coastal erosion, desertification;
- Presence of regional structures and/or processes for coastal management to increase the certainty of success of the project;
- Territorial continuity with other selected regions to test procedures and structures for an effective coordination of areas shared by different Regions.

The Feasibility Study (FS), carried out by Regions involved in the project, revealed a complex scenario that hinders the integrated management of coastal areas, mainly because of the fragmentation existing among the different levels of government. In particular, the main problems in the CAMP Areas, as identified in the FS, are related to the absence of a centralized responsible authority for the management of coastal areas. In fact, while acknowledging the IMELS a focal role in the field of integrated management of coastal areas, nowadays Italy has not a coastal Act neither a dedicated coastal governance body or management Agency. Coastal management is carried out within a range of regulatory frameworks dealing with national government and regional levels related to administrative decentralization sectors such as environment, agriculture, fisheries, industry, infrastructure and transport, tourism, education and research, health, economic development. This result is partly due to the process of decentralization of certain matters from the State to the Regions.

The CAMP Italy Project was conceived to promote the relationships among the coastal regions involved in it, improving the existing networks, and strengthening communication and arrangements for the exchange of experiences. In particular, the regional administrations, through the CAMP Italy, have the opportunity to meet and exchange views on different approaches to Integrated Coastal Zone Management. Other regions, although not directly involved in CAMP Italy, may take profit of the outcomes of the Project through the planned CAMP information and dissemination activities. The ability to analyse issues and to program the management along with possible solutions, offers a great opportunity for growth in regional institutions, thanks to the fact that, in CAMP Italy, different regional choices are examined and compared and, through an effective coordination, management plans providing the best solutions can be reached. At the same time, through the actions taken by each individual Region, it is possible to obtain and evaluate the results which will enable other Regions to adopt and/or modify the project solutions, according to their own needs. In summary, the implementation of the CAMP Italy can be justified by the following reasons:

- Diversity and representativeness of values and criticalities (environmental, cultural and socio-economic) compared to the national context and the Mediterranean basin;
- Territorial diversity expressed in terms of environmental, cultural and socio-economic aspects as a prerequisite for a high degree of intensity and complexity of multidimensional relationships;
- Information flow and shared project guidelines among CAMP stakeholders;
- Adoption of "bottom-up" and "step-by-step" strategies;
- Promotion of project initiatives that consolidate some experiences of inter-institutional coordination already in place.

The chapter describes the CAMP Italy project and present the adopted solutions to apply and achieve ICZM at both Italian national and local level.

1.2. Objectives

The CAMP Italy Project aims to incorporate the principles of sustainable development and integrated coastal zone management in the development process to better articulate marine and coastal resources protection and development objectives. A major goal of the CAMP Italy Project is to encourage the implementation of national strategies for integrated coastal zone management. In accordance with the implementation of the ICZM Protocol in the Mediterranean, the CAMP Italy Project helps to integrate, standardize, share and implement tools and procedures aimed at adapting the territorial planning, with implications for the management of coastal resources and assets. The CAMP Italy Project also aims at favouring integrated coastal zone management in Italy, not only by selecting areas that may represent the whole national situation, but also by promoting, in some of these areas, an agreed governance among the local, regional and central administrations. Through the

CAMP Italy actions the involved administrations can take profit of an additional tool for the protection of coastal ecosystems, for inter-institutional coordination, and for becoming part of an international community, thus improving professional growth and policies aimed at environmental conservation and management. Table 1 lists all objectives of the CAMP Italy project.

Table 1. Obiectives of the CAMP Italy project

One Strategic objective
Testing the integrated management of the coastal area, implementing both the ICZM Protocol and the EU ICZM Recommendation 2002
Three specific objectives
1. Improve the coordination capacities among different structures involved in coastal zone management (governance for coastal area) and both vertical (network of actors) and horizontal (industry sectors) integration. 2. Maintenance of the ecosystem service flow through the reduction and control of human pressures and development of adaptation to the effects of climate change. 3. Verification of the sustainability of socio-economic activities which stress the coastal area through application of the ecosystem and the economic evaluation of ecosystem services (PES, TEEB, etc.); in particular tourism, fishing, coastal defence.
Seven complementary objectives
1. management of the marine spaces (such as management of the mining of sand from the seabed, of fish farms, etc.); 2. protection and enhancement of the historical-cultural and environmental heritage, and landscape; 3. diversification of the touristic offer (new offers focused on specific environmental and landscape aspects); 4. environmental and landscaping restoration; 5. recovery and preservation of the coastal and marine environment at the river mouths; 6. description and representation of the territory and its resources, identifying values, opportunities and critical issues, reasons for the proposed choices and evaluating expected effects; 7. definition of evolutionary scenarios which intervention policies must address.

1.3. The Three Regions

The three selected Regions part of the CAMP Italy project are: Emilia-Romagna, Tuscany and Sardinia.

Tuscany is characterized by an important archipelago formed by a group of seven islands that constitute the National Park (Tuscany Archipelago National Park) and it is the largest Marine Park in the Mediterranean, within the Pelagos Sanctuary for Mediterranean Marine Mammals.

Emilia-Romagna is characterized by the massive presence of tourism and related activities. It is an area in which the coastal governance has already established tools (cognitive, monitoring, planning and legislative) for developing coastal sustainable management. The area selected by the CAMP Italy project in Emilia-Romagna encompasses environmental values and human activities of significant socio-economic importance. In fact, the pilot area belongs to the Po Delta Regional Park, which since its foundation works to identify solutions to make compatible the instances of conservation and environmental protection with the economic and productive aspects and needs of the residents, thus combining the conservation issues with socioeconomic development. The Region has already developed several projects on Integrated Coastal Zone Management: in particular a project for the definition of the ICZM guidelines and a program of experimental actions sponsored by the Po Delta Park (both in 2005).

Sardinia can be considered as a complete and isolated coastal system in the middle of the Mediterranean Sea, and it offers the opportunity for spatial planning and sustainable development of coastal areas not yet affected by the presence of human settlements and activities. The Autonomous Region of Sardinia developed a specific governance tools for integrated coastal zone management: the "Regional Coastal Conservatory of Sardinia"- RCCS, a Sardinian public Agency created in 2007 in order to ensure the protection of outstanding natural areas on the Sardinian coast and to respond appropriately to instances from different sectors and local governments. The RCCS has specific functions related to the coordination of initiatives for integrated management of coastal zones carried out by Regional Administration, Local Authorities, the Administrations of Natural Protected Areas, and EU Natura 2000 sites and, in addition, to the promotion and dissemination of issues related to environmental protection, landscape and sustainable development in coastal areas.

2. Methodology of the CAMP Italy Project

The approach chosen for the design and implementation of the CAMP Italy Project recognizes and incorporates the following fundamental considerations:

- CAMP Italy is a "country-driven" project. The Project which is designed to respond to national, regional and local priorities. It operates according to the experiences and realities of the Italian legal, administrative and cultural institutions and uses the available national and local expertises;

- CAMP Italy Project is based on an integrated approach to marine and coastal environment and development issues. Project activities cut across protection and development problems, harmonize public sector priorities and private sector pressures, and provide an integrated strategy for the common goal of achieving development in the region within a sustainable management policy framework.

The Project strategy is based on Integrated Coastal Zone Management (ICZM) and on the Ecosystem based Approach (EcAp) as a key tools for planning sustainable development based on sustainable development principles. ICZM and EcAp refer to the relevant International, European Union and Domestic Law and related plans and programs, in particular according to the principles and objectives set by the 2008 ICZM Protocol of the Barcelona Convention. Moreover the project strategy refers to the relevant EU legislation, such as the new Directive 2014/89/EU establishing a framework for Maritime Spatial Planning; the Marine Strategy Framework Directive 2008/56/EC; the Habitat Directive 92/43/EEC and the Birds Directive 2009/147/EC; the Regulation 508/2014/EU on the European Maritime and Fisheries Fund; the Recommendation 2002/413/EC concerning the implementation of ICZM in Europe; the Decision 2010/631/EU concerning the conclusion on behalf of the European Union of the ICZM Protocol; the Directive 2009/28/EC on the promotion of the use of energy from renewable sources; the Regulation (EC) No 2371/2002 on the conservation and sustainable exploitation of fisheries resources under the Common Fisheries Policy; the new Common Fishery Policy (Regulation No. 1380/2013); the Directive 2000/60/EC establishing a framework for Community action in the field of water policy; the Directive 2003/35/EC providing for public participation in respect of the drawing up of certain plans and programmes relating to the environment.

Project phasing follows the approach incorporated in the ICZM project planning process, consistently with the manual on the preparation and implementation of the CAMP programme (UNEP/MAP, 1999). Phases, stages and outputs are reported in Figure 2.

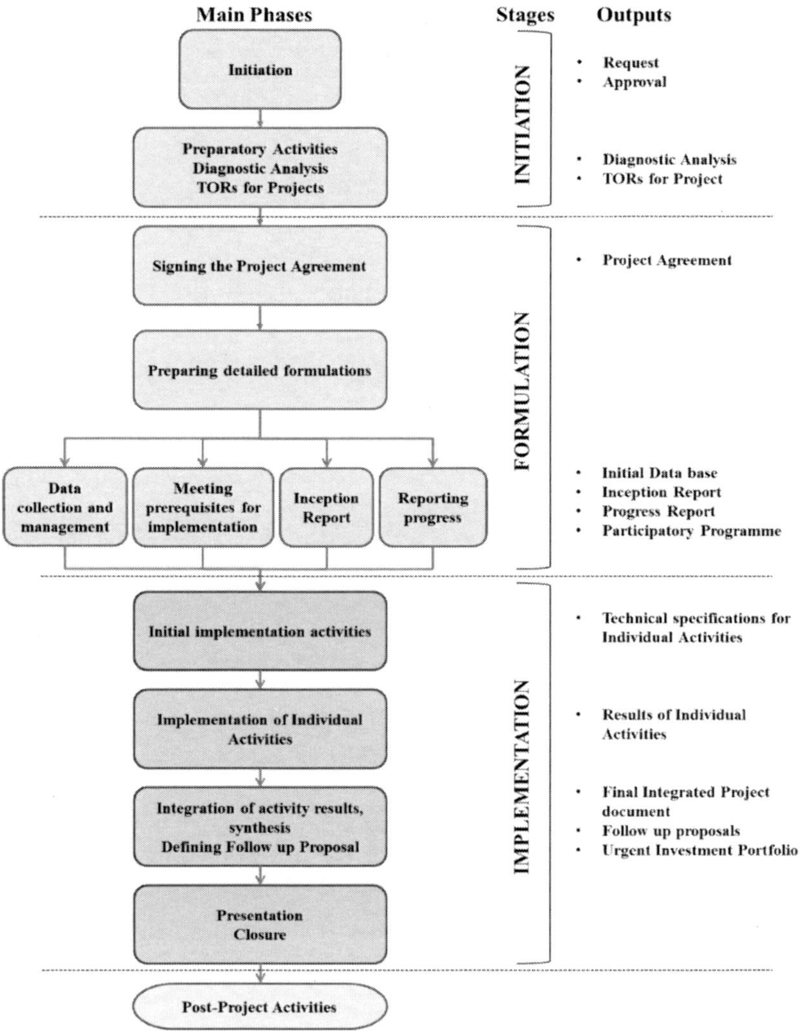

Figure 2. Phases, stages and outputs (modified from UNEP-MAP, 1999).

The CAMP Italy project is structured in two main segments:

- *Horizontal Activities:* based on data, problems, discussions, documents aimed at four goals: I. coordination and dissemination of the results; II. capacity building; III. data collection and management; and IV. institutional coordination and public participation, which are discussed in the following paragraphs;

Table 2. Individual Activities developed within the CAMP Italy project. Each activities is related to the specific objectives and Thematic Area of the Project

Specific objectives and corresponding Thematic Area	Individual activities
Objective 1: Improve the coordination capacities among different structures involved in coastal zone management (governance for coastal area) and both vertical (network of subjects) and horizontal (industry sectors) integration. **Thematic Area 1:** Planning of marine and terrestrial coastal areas	**Emilia-Romagna** (i) Regulations for the removal of sand and material from marine quarries; (ii) Operational protocol for the reestablishment and consolidation of the dune belts (natural or artificial/reconstructed); (iii) Education and communication to sustainability. **Sardinia** (i) Sardinian Center for coastal monitoring and assessment; (ii) Coastal vulnerability assessment tool; (iii) Implementation of specific actions aiming to the sustainable use of beaches and the protection of coastal zones; (iv) Sardinian Center for coastal monitoring and assessment; (v) Coastal vulnerability assessment tool. **Tuscany** (i) Monitoring the effects of Tuscany beach nourishment activities in the South of Ansedonia.
Objective 2: Maintenance of the ecosystem service flow through the reduction and control of human pressures and development of adaptation to the effects of climate change. **Thematic Area 2:** Protection, preservation and restoration of coastal and marine habitats.	**Emilia-Romagna** (i) Conservation of marine and coastal biodiversity with special reference to the conservation of minor species; (ii) Education and Communication to sustainability. **Sardinia** (i) Asinara National Park and Pianosa Island (Tuscany Archipelago National Park); (ii) Coastal vulnerability assessment tool **Tuscany** (i) Evaluating feasibility and effectiveness of the use of alluvial sediments extracted from the Magra River drainage basin in order to restore eroding beaches in the annexed coastal cell; (ii) Protection of the dunes of Lacona (Elba Island).
Objective 3: Verification of the sustainability of socio-economic activities which stress the coastal area through application of the ecosystem and the economic evaluation of ecosystem services; in particular tourism, fishing, coastal defence. **Thematic Area 3:** Sustainability of socio-economic stress on the coastal zone (in particular tourism, fishery, structures for coastal defense).	**Emilia-Romagna** (i) Development and exploitation of new forms of sustainable tourism in protected natural areas. **Sardinia** (i) GIRA – Integrated Fishing Resources Management: (ii) Asinara National Park and Pianosa Island (Tuscany Archipelago National Park) tourism Carrying Capacity Assessment; (iii) Enhancement of architectural heritage for the creation of the first eco-hostel in the coastal area for sustainable local development **Tuscany** i) Asinara National Park and Pianosa Island (Tuscany Archipelago National Park) tourism Carrying Capacity Assessment

- *Individual Activities* structured around the main thematic areas: I. Planning of marine and terrestrial coastal areas; II. Protection, preservation and restoration of coastal and marine habitats; and III. Sustainability of socio-economic stress on the coastal zone (in particular tourism, fishery, structures for coastal defence).

The Horizontal Activities are those actions, expressly provided for by UNEP/MAP, which characterize a CAMP Project. They can be regarded as mandatory and therefore they have to be included in any CAMP Project. The Individual Activities are those actions carried out by the involved Regions. The Individual Activities were defined by the Project Steering Committee with the aim to address the specific objectives of the project. Each Region was asked to develop at least one activity to address the three thematic areas (Table 2).

3. Results

In the following sections, results achieved so far in the implementation of the CAMP Project are outlined and described. It has to be noted that the Project is expected to end in December 2016 and during the writing of this chapter, some individual activities have not started yet and others are ongoing. A full analysis of Individual Activites and related results was not possible and in this section the focus is mainly on the Horizontal Activities.

3.1. Coordination and Dissemination of the Results

In the field of ICZM, there is a need for closer coordination between national authorities and regional and local bodies in order to provide a more effective evaluation also in relation to the variuos authorizations for activities that may influence coastal zones. Therefore, adequate joint consultative bodies or joint decision-making procedures are discussed and "experimental" institutional coordination (that interfaces consultative bodies, inquiries or public hearings procedures and possible partnerships) is carried out during specific meetings and workshops. The results are thus used to measure the efficacy of identified ICZM strategies and approaches in order to discuss the elaborated ICZM instruments and related weaknesses and strengths for the governance applied to specific issues, such as maritime/marine and coastal spatial planning and management. Other initiatives were planned, playing a major role in presenting, promoting and disseminating project results and ICZM issues, either in the framework of CAMP Italy conferences or thanks to the participation to other events of dissemination. Lessons learned from Individual Activities, at the end of the implementation, will be carefully analysed and considered as the basis for new proposals regarding ICZM implementation at national and regional levels that will be presented during the Final Conference (December 2016). The comparative analysis of the Individual Activities will allow to identify the constraints and administrative problems that may hinder future actions in the field of Maritime Spatial Planning (MSP) and ICZM. It will therefore lead to the definition of shared procedures for the resolution of conflicts, mitigation of impacts, environmental restoration/conservation, land use planning, etc.

3.2. Capacity Building

The overall objective of this Horizontal Activity is to build and strengthen skills and experiences, while, in accordance with article 15.2 of the ICZM Protocol, guarantee the training and improving competencies of public Administrations at national, regional and local scale participating in the CAMP Italy Project. This goal is pursued primarily by designing and implementing innovative ways to maximize the effectiveness of the strategies outlined in documents issued by UNEP, as well as by EU on policies related to the integrated management of the coastal zone. In particular, the specific objectives of this Horizontal Activity are:

- Developing adequate planning skills in order to be compliant with relevant references about integrated management of the coastal zone;
- Pointing out potential criticalities in decision-supporting and decision making;
- Defining and suggesting opportunities for improvement and/or corrective actions;
- Promoting institutional development (leveraging, whenever needed, the involvement of stakeholders and other relevant societal components), the development of human resources and the strengthening of management systems.

Capacity building favours an overall vision shared between the different Authorities (at national, regional and local scale) about possible planning activities aimed at both marine and terrestrial coastal areas. Lessons learned from Individual Activities are used as a basis for highlighting the need for a tighter integration between marine and terrestrial coastal management. Awareness of problems and opportunities at different scales and organizational levels certainly prove to be a key to an effective and consistent implementation of integrated management in coastal areas.

Capacity building activities take place in CAMP Italy Regions as well as at national level. The identification of potential participants to capacity building activities takes place on the basis of recommendations from the heads of the various Administrations involved (e.g.,, Regions, Provinces, Municipalities; Basin Authorities, Park Authorities, Port Authorities, and others who have expertise on coastal zone planning).

Before starting the capacity building activities, a questionnaire aimed at defining the level of knowledge about ICZM was delivered to the participants (Questionnaire 1). The questionnaire allowed to define not only the starting level of awareness about ICZM issues, but also to optimize the selection of the topics to be presented and discussed. At the end of the capacity building activities, a new questionnaire (Questionnaire 2) is delivered as well. This questionnaire, in addition to assessing the final level of knowledge about ICZM issues, allows to collect hints and suggestions about the best corrective actions with respect to the issues that participants had previously indicated in Questionnaire 1.

Moreover, e-learning models as training courses have been structured and carried out with the cooperation and technical support of other project partners and INFO/RAC in order to create a common basis of knowledge among the actors involved.

In addition, the Capacity building activity include the following:

- Creation and/or improvement of regions networking opportunities;
- National and international MSP and ICZM consultant connection.

Capacity building activities are carried out by staff of IMELS, regional and governmental agencies (either involved or not in the CAMP Italy Project, e.g.,, municipalities, park authority, etc.), as well as by PAP/RAC, other RACs, and by external consultants under the guidance of the National Project Coordinator. The possibility to organize a MedOpen virtual training course on ICZM is foreseen by PAP/RAC for participating regions. The main tool that is used to this purpose is the organisation of workshops in which the different actors are involved in discussions and analysis of the main environmental issues and their integrated management. The expected outputs from this horizontal activity are:

- Information and training to decision makers on planning projects in the coastal zone;
- Development and improvement of performances and operating procedures of the bodies involved in the ICZM;
- Creation and/or improvement of networking;
- Endorsement of experts' connections.

3.3. Data Collection and Management

Information about Italian coastal zones is available from many different sources, ranging from those focusing on small scale features (e.g., cadastral maps) usually managed by local Administrations, to those focusing on large or global scale thematic layers that are usually freely available online on local geoportals. Unfortunately, sometimes information is outdated, while different data sources are updated at different times, making synoptic data analyses very difficult, if not impossible. The same problems affect marine data sources. Marine data that are potentially useful to coastal zone management are in most cases very sparse in time and space, and data sources are seldom synchronous.

This Horizontal Activity does not solve all the information-related problems that may hinder Italian coastal zone management, but aims at defining guidelines for effective data interoperability between CAMP Italy regions and for data sharing at national level. As the assessment of the ecological status of marine waters plays a major role in recent EU directives (e.g., the Water Framework Directive -Dir. 2000/60/EC- or the Marine Strategy Framework Directive -Dir. 2008/56/EC-), information about marine ecosystems are carefully taken into account, especially in line with the need for a better integration with terrestrial data sources (e.g., those related to Habitats -Council Dir. 92/43/EEC-, Birds -Dir. 2009/147/CE-, Environmental Quality Standards -Dir. 2008/105/EC-, and Bathing Water Directives -Dir. 2006/7/EC-) and with other relevant EU legislation (including the Common Fisheries Policy -Regulation EU No. 1380/2013-). Similar attention is paid to economic and societal analyses regarding the use of marine ecosystems and the cost of their degradation, which are expected to play a very relevant role in the future, although currently they are still in their earliest stages.

Retrieving, collecting, managing and analysing the available information are very common tasks in supporting decision making. Effective user-friendly tools for managing complex scenarios in the most effective way can be developed. The ICZM Protocol pointed out the need for collecting updated information on resources, activities, institutions, legislation and planning as a preliminary stage of any sensible action to be carried out in the coastal zone. The "Data collection and management Activity" is obviously aimed at favouring this process as far as possible, but its ultimate goal is to define effective guidelines for data collection and management optimization, with special reference to data interoperability and harmonization. In fact, CAMP Italy is trying to harmonize spatial planning data and related metadata among the participating Administrations, taking into account the European INSPIRE legislation (Dir. 2007/2/EC and following Regulations) and international standard and aiming at defining the basic requirements for a Spatial Data Infrastructure (SDI). The SDI implements a framework of geographic data, metadata and tools that can be easily and interactively connected, thus making any use of spatial data as efficient and flexible as possible. The SDI should encompass not only technological issues or data standards, but also policies, institutional arrangements and human resources that make the available information accessible by the widest range of potential users. Geographic Information Systems (GIS) should play a major role by underlining the necessity to use and develop an interoperable open source system for collection and harmonization of data among national and Mediterranean partners, in order to monitor coastal vulnerability and implement specific data on ICZM. While defining the guidelines for the implementation of an SDI is the main goal of this Horizontal Activity, defining a shared set of thematic information layers is a secondary, but not less important, goal. This activity includes the following actions:

- Overview and collection of the data sets produced by the CAMP Regions to support the Individual Activities;
- Development of a spatial data infrastructure (Geo-WebService) that aims to facilitate spatial data use, integration and sharing in compliance with INSPIRE guidelines;
- Definition of the largest subset of thematic information layers which are available in all the CAMP Regions and upload them on the project Web-GIS;

- Identification of a set of other sources of information that may support CAMP activities and, from a more general viewpoint, future MSP/ICZM activities and integration of this information on the project Web-GIS;
- Definition of possible data conversion procedure, if data formats are not homogenous among Regions;
- Editing of a Data Policy document, which identifies the rules of data use, as well as the access to information and products generated within the activities of the Project. Specifically, the Data Policy proposal aims to encourage as much as possible the rapid sharing of information and data, in line with the provisions of the ICZM Protocol (Art. 16) and, primarily, with the European Directives adopted by Italian law (i.e., the Directive on the re-use of public sector information, also known as PSI Directive - public sector information - Dir. 2003/98/EC-, as amended by Dir. 2013/37/EU of 26 June 2013; the Law 135/2012 "Spending review," art. 12 quaterdecies), which promotes the full usability of the data coming from public funds;
- Collection of data about results and durability of Individual Activities, to support the assessments activities;
- Guidelines for achieving a full interoperability of the shared SDI.

Along with these activities, another useful tool is developed and adopted by the project to support the implementation of ICZM principles: the dataset of indicators. The database includes a total of 49 indicators (environmental, social, economic and governance indicators), which have been selected after an accurate review of the available indicators developed in previous national and international projects: Pegasus Project (http://www.pegasoproject.eu/); SUSTAIN Project (http://www.sustain-eu.net/index.htm); Recap-ICZM Project (UNEP/MAP http://www. unepmap.org/); Progetto Coste (Coast Project) (http://www. pcn.minambiente.it/GN/progetto-coste); Guidelines for Mediterranean Marine Protected Areas (MPA) management (https://portals.iucn.org/library/ node/10406). The indicators are applied to the 5 CAMP areas during the Individual Activities performed by the Regions. The final calculation of the indicators is developed by the regions and the results will be implemented on the abovementioned geographic platform (Web-GIS).

The indicator database is published online on the project website (http://www.camp-italy.org/) and is accessible by all project partners who can update, modify the information, thus making the database an useful tools for evaluating sustainability and measuring the implementation of ICZM principles into the individual tasks.

3.4. Institutional Coordination and Public Participation

This activity aims at promoting vertical integration and the collaboration between Administrations for a sound coastal management, and to ensure appropriate participation in future decision-making and spatial planning processes. Specific objectives are:

- Promoting cross-sectorial institutional coordination among the various administrative services and regional and local authorities competent in coastal zones, with the ultimate aim of facilitating the creation of an ICZM national network;
- Ensuring the adequate and timely participation in the transparent decision-making process by local populations and stakeholders (e.g., private and public workers associations, environmental groups, NGOs, general public).

The first specific objective is functional to create and test the National ICZM network which has been involved in the CAMP activities set out, among others, in Article 32 (Institutional Coordination) of the ICZM Protocol. Ensuring institutional coordination among the various authorities (national, regional and local) competent for both marine and land parts of coastal zones is mandatory to strengthen the coherence and effectiveness of the coastal strategies, plans and programs.

The second general objective is functional to elaborate and testing the adequate instruments, procedures and measures specifically dedicated to ensure the appropriate involvement of territorial communities and public entities, economic operators, non-governmental organizations, social actors and the public concerned in the phases of the formulation and implementation of coastal and marine strategies, plans and programmes or projects, as indicated in Article 14 of the ICZM Protocol (Participation). Thus, it is underlined the necessity of developing an institutional framework on ICZM at the national level, while mapping stakeholders, testing models of participation, and informing decision-makers. Participation is an important feature in ICZM projects and CAMP projects. In CAMP projects, the participation of all stakeholders should be provided at all stages, including the design phase.

Therefore, a participation strategy has been developed according to the involved institutional actors and other stakeholders priorities and to the level at which they may be involved: low level of public involvement or influence (inform, educate, share or disseminate information); mid-level (gather information, views); high level of public involvement or influence (discuss through two-way dialogue); very high level of public involvement or influence (fully engage on complex issues, partners are involved in the implementation of solutions). At Regional level several participation instruments are activated: I. informing the public through events dedicated to provide citizens of the different coastal municipalities with the fundamental principles and objective of CAMP Italy; II. involving the public by focus group dedicated to the actual involvement of stakeholders in the analysis and discussion of CAMP Italy results and outputs; III. stimulating and promoting the structuring and functioning of Partnerships of two or more organizations (between local authorities and multi-disciplinary academic experts; among homogenous entities (i.e., fishermen, tourist operators, islands populations, etc.) co-operating as a unit to influence developments for their common good, perform effective lobbying, improve exchange of experiences and ideas, enhance abilities to develop projects, etc).

This activity also offers the opportunity to run pilot experiences to test the appropriate governance allowing participation, such as: I. verifying how to facilitate public participation to CAMP Italy in particular and to ICZM process in general, through new communication technologies (e.g., mobile data collection); II. setting of advisory committees or steering groups in a voluntary way; III. analysing the procedures followed where public consultation is mandatory (e.g., Environmental Impact Assessment, Strategic Impact Assessment, spatial planning procedures).

The objectives of the participation strategy can be summarized as follows: I. protecting and disseminating comprehensive and transparent information; II. informing key stakeholders about project activities; III. building partnerships between the public, private entities, associations, and the scientific community; IV. contributing to the resolution of conflicts and reaching consensusc.

As part of the CAMP Italy project, a virtual participation tool is defined through the www.joincampitaly.org platform to:

- Creating a web space that encourages citizen participation;
- Giving visibility to issues related to the integrated management of coastal areas by ensuring a space for discussion between users interested in the topics of discussion;

- Improving transparency in the activities related to ICZM carried out by the Partner regions in the respective pilot areas.

Joincampitaly.org is conceived as an open, interactive and constantly updated virtual space. The platform is designed to enable all persons, interested in the issues of coastal zone management, to actively participate in the design of the future of the coastal zones. The website contains the description of the activities, including the state of play, the preparatory documents, surveys, topics for discussion and the open discussions. The portal is also the place which virtually store and revitalize the themes discussed and developed during public meetings or workshops or any other form of public participation.

In particular the portal counts on three main tools to stimulate and engage citizens, administrators and all users interested in the topics regarding the coastal zones:

- Forum: It is the virtual space dedicated to the discussion of the three strategic fields of CAMP-Italy project (the three Thematic Areas, see Table 2). There is a new discussion topic each month for a total of nine months (3 discussion topics by subject area);
- Blog: For each Individual Activity the website offers a short description and the updated implementation status. Discussions regarding the Individual Activities are animated and facilitated by the regions;

Surveys: They have the main scope of involving the general public on current issues regarding coastal zones. Any user can access the platform for consultation without signing in, but in order to actively participate and be able to give their own contributions, users must register to the site and create a personal account. The only exception to this rule, are the surveys that are open to all visitors without any registration.

3.5. Communication

This Activity provides the elaboration of a communication strategy, at both Regional (e.g., CAMP Regions) and National levels, outlined in the communication plan, assuring the application of Article 15 of the ICZM protocol - which requires to carry out awareness-raising activities on integrated coastal zone management and to develop educational programmes, training and public education on this subject. The main objective of the Communication Plan is to provide a general communication strategy for the CAMP project facing the complexity of the project itself (14 pilot actions carried out in 5 Areas within 3 Regions). The communication objective are the following: I. raising the awareness about ICZM; II. supporting public participation; III. supporting the capacity building process; IV. improving a network of stakeholders; V. spreading the CAMP Italy results at National and Mediterranean level.

In order to provide an adequate communication strategy, targets groups have been clearly identified and grouped in: general public, public institutions and public services, economical actors and experts.

The communication tools and channels are:

- Web site which collects all the informations about the project;
- Public events such as workshop, mid term and final conference;
- Video productions that includes two short video productions and the video report of a conference/workshops;
- E-Newsletter;
- Social Media including facebook page and youtube channel;
- Forum and blog for participatory process;
- Press releases for the media.

For each target group a set of key messages has been identified. For the general pubblic the key messages focuses on coastal zones as public good, on the importance of regulating the human presence in coastal zones and on the importance of ecosystem services and fragility of coastal ecosystems.

For economical actors contents underline the importance of economical activities based on the coast, on the scarcity of resources, on the reliance of economical activieties on ecosystem services and on the importance of institutions and public/private dialogue.

For Institutions and public bodies messages strengthen the importance of ICZM as a tool to improve coastal management, on the complexity of coastal zones, on the necessity to involve citizens and economic actoris in coastal management processes.

The key messages for Experts underline the importance of building a network of “ICZM minded” professionals to support decision making. Each message has been associated to one or more communication tools: this helps keeping communication activities under control (covering all the targets with all the message). In order to monitor communication activities, a set of indicators for each tool has been provided.

CONCLUSION

Today, the impact of human activities increases the exposure of coastal zones and the risk of changes. Scientists confirm that these risks are shared by all Countries and are increasingly demanding for the advancement of knowledge and joint efforts for a better understanding of the adaptation approaches. Building on the uniform and consistent application and interpretation of existing legislation, describing the new developments in the field of ICZM can and should help to substantially improve the current conditions of coastal zones and their ecosystems through shared management strategy for implementation comprehensive plans and programs. ICZM is a way of operating that can meet this crucial challenge for the future of coastal areas in general and particularly for the fragile and vulnerable coastal zones. Currently ICZM is strongly supported by public policies at all levels of governance (from local to national and even international) in order to lead coastal areas to sustainable development, which is the real guarantee of progress and of acceptable quality of life. The various initiatives undertaken in the Mediterranean are important promoters of a sustainable development of the European coast in general and the Mediterranean in particular.

The CAMP Italy Project is the first multi-area project within the CAMP programme. Through the acquisition and sharing of knowledge and experiences, the understanding of critical coastal areas, the CAMP Italy constitutes an essential tool for the consolidation of ICZM and the so called 'land-sea interactions' in Italy. The fragmentation existing among the different levels of the Italian government has been one of the main issue for the application of the principles of the ICZM Protocol in the past, along with he absence of a centralized responsible authority for the management of coastal areas. Through the involvement of all these levels and the relevant actors, the CAMP Italy Project allows to strengthen the land-sea management policies and environmental integration in the management of coastal areas in Italy and can be a model for the rest of the Mediterranean area. The commitment to a CAMP project, which has only a very limited period in time (within 2 years) should be considered as a way to boost, to accelerate or to initialize a new way to manage coastal zones. This implies on the part of the IMELS a constant concern in view of the post-CAMP phase, concrete follow up (ICZM plan or strategy), but also benefits for the entire Italian coastal zone. Its recognition at the national level is undeniable. This indicates that a significant additional effort is needed to highlight the achievements and draw lessons from the CAMP exercise. In particular, the assessment on how the issue of management coastal area and the practice of ICZM were processed is needed. Whereas the conduct of CAMP was a learning for all, this analysis is essential for the sustainability of the ICZM processes activated. The factors affecting the sustainability of the ICZM have been discussed and evaluated by the Steering Committee ("Tavolo CAMP") during the CAMP implementation.

Run a CAMP project represents both a technical and strategic operation for its initiators, PAP / RAC and IMELS. The CAMP Italy, and the multi-area approach, cannot waive the rule and must, instead, refer to one of the most successful realization among the CAMP projects carried out so far in the Mediterranean. Indeed, it should not go without noting that the implementation of the CAMP has helped to highlight: firstly, the validity of the methodology supported by PAP / RAC that addresses all the ins and outs of the ICZM process by promoting the principles, tools and methods that are fully dedicated to him. This methodological framework has a structuring character for the regional administrations and for other stakeholders very valuable and useful; on the other hand, some of applicability defects that can cause difficulties in its implementation. The adaptable nature of the ICZM Protocol is therefore necessary, taking into account the particularities and specific peculiarities of the coastal areas.

Acknowledgments

The authors gratefully acknowledge the Italian Ministry of the Environment and Protection of Land and Sea of Italy, Emilia-Romagna, Tuscany and Sardinia Regions, and the Priority Actions Programme Regional Activity Centre for co-funding the CAMP Italy Project. The authors also thanks the MAP Components involved in the project, such as the Regional Activity Centre for Information and Communication (INFO/RAC), the Plan Bleu (BP/RAC), the Regional Activity Centre for Specially Protected Areas (RAC/SPA), the Regional Activity Centre for Sustainable Consumption and Production (SCP/RAC) and MED POL.

Project website: http://www.camp-italy.org/

REFERENCES

Council Decision 2010/631/EU of 13 September 2010 concerning the conclusion, on behalf of the European Union, of the Protocol on Integrated Coastal Zone Management in the Mediterranean to the Convention for the Protection of the Marine Environment and the Coastal Region of the Mediterranean.

Council Directive 92/43/EEC of 21 May 1992 on the conservation of natural habitats and of wild fauna and flora; *Offical Journal*, L206, 22/07/1992; pp. 7-50.

Directive 2000/60/EC of the European Parliament and of the Council of 23 October 2000 establishing a framework for Community action in the field of water policy; *Offical Journal*, L327, 22/12/2000, pp. 1-73.

Directive 2003/35/EC of the European Parliament and of the Council of 26 May 2003 providing for public participation in respect of the drawing up of certain plans and programmes relating to the environment and amending with regards to public participation and access to justice Council Directives 85/337/EEC and 96/61/EC; *Offical Journal*, L156, 25/06/2003, pp. 17-25.

Directive 2003/98/EC of the European Parlimanet and of the Council of 17 Novemeber 2003 on the re-use of public sector information; *Official Journal*, L345, 21/12/2003, pp. 90-96.

Directive 2006/7/EC of the European Parliament and of the Council of 15 February 2006 concerning the management of bathing water quality and repealing Directive 76/160/EEC; *Offical Journal*, L64, 4/3/2006, pp. 37-51.

Directive 2007/2/EC of the European Parliament and of the Council of 14 March 2007 establishing and Infrastructure for Spatial Information in the European Community (INSPIRE); *Offical Journal*, L108, 25/4/2007, pp. 1-14.

Directive 2008/105/EC of the European Parliament and of the Council of 16 December 2008 on environmental quality standard in the fields of water policy, amending and subsequently repealing Council Directives 82/176/EEC, 83/513/EEC, 84/156/EEC, 84/491EEC, 86/280/EEC and amending Sirective 2000/60/EC of the European Parliament and of the Council; *Offical Journal*, L348, 24/12/2008, pp. 84-97.

Directive 2008/56/EC of the European Parliament and of the Council of 17 June 2008 establishing a framework for Community action in the field of marine environmental policy (Marine Strategy Framework Directive); *Offical Journal*, L164, 25/06/2008, pp. 19-40.

Directive 2009/28/EC of the European Parliament and of the Council of 23 April 2009 on the promotion of the use of energy from renewable sources and amending and subsequently repealing Directives 2001/77/EC and 2003/30/EC; *Offical Journal*, L140, 5/06/2009, pp. 16-62.

Directive 2009/147/EC of the European Parliament and of the Council of 30 November 2009 on the conservation of wild birds; *Offical Journal*, L20, 26/1/2010, pp. 7-25.

Directive 2013/37/EC of the European Parliament and of the Council of 26 June 2013 amending Directive 2003/98/EC on the re-use of public sector information; *Offical Journal*, L175, 27/6/2013, pp. 1-8.

Directive 2014/89/EU of the European Parliament and of the Council of 23 July 2014 establishing a framework for maritime spatial planning; *Offical Journal*, L257, 28/8/2014, pp. 135-145.

Law 135/2012 of the Italian Counsel Of Ministry of 7 July 2012 - Spending review.

Protocol on Integrated Coastal Zone Management in the Mediterranean (2008), *Official Journal*, L034, 04/02/2009, pp. 19-28.

Recommendation 2002/413/EC of the European Parliament and of the Council of 30 May 2002 concerning the implementation of Integrated Coastal Zone Management in Europe, (OJ L 148, 6.6.2002, p. 24).

Regulation (EC) No 2371/2002 of 20 Decemebr 2002 on the conservation and sustainable exploitation of fisheries resources under the Common Fisheries Policy; *Offical Journal*, L358, 31/12/2002, pp. 59-80.

Regulation (EU) No.508/2014 of the European Parliament and of the Council of 15 May 2014 on the European Maritime and Fisheries Fund and repealing Council Regulation (EC) No. 2328/2003, (EC) No. 861/2006, (EC) No. 1198/2006 and (EC) No. 791/2007 and Regulation (EU) No. 1255/2011 of the European Parliament and of the Council; *Offical Journal*, L149, 20/05/2014, pp. 1-66.

Regulation (EU) No. 1380/2013 of the European Parliament and of the Council of 11 December 2013 on the Common Fisheries Policy, amending Council Regulations (EC) No. 1954/2003 and (EC) No. 1224/2009 and repealing Council Regulations (EC) No. 2371/2002 and (EC) No. 639/2004 and Council Decision 2004/585/EC; *Offical Journal*, L354, 28/12/2013, pp. 22-61.

UNEP/MAP (1999) Formulation and Implementation of CAMP Projects: *Operational Manual*, *MAP-PAP/RAC*, *Athens – Split.*, ISBN 953-6429-23-3.

In: Coastal Zones
Editor: Gina Moran

ISBN: 978-1-63485-611-9

Chapter 2

ASSESSMENT OF COASTAL WETLANDS IN LEBANON

Amin Shaban[1,*], Ghaleb Faour[2], Raya Stephan[3], Carla Khater[1], Talal Darwich[1] and Mouin Hamze[1]

[1] Lebanese National Council for Scientific Research, CNRS, Beirut, Lebanon

[2] Remote Sensing Center at the Lebanese National Council for Scientific Research, CNRS, Beirut, Lebanon

[3] International Consultant, France

ABSTRACT

The unique landscape, where wetlands are located, needs to be monitored and reserved, since their continuity depends on the balanced input-output water system. The hydrologic setting of wetlands is always creating unique ecosystems for flora (e.g., pollens, etc.) and fauna (e.g., migratory birds, etc.). In Lebanon, wetlands are well pronounced, and they own different dimensions ranging from several hundreds of square meters up to few square kilometers. Lately, the Lebanon's wetlands have been subjected to physical and anthropogenic stresses, notably the water feeding becomes crucial, and this makes some wetlands losing their hydrologic criteria (i.e., water-saturated terrain). The changing climatic

* Corresponding author: Dr. Amin Shaban. Field of specialization: Hydrology, Water Resources Management and rRmote Sensing. Address: Remote Sensing Center, CNRS, Zahia Salman St., P.O. Box 11-8281, Beirut, Lebanon. E-mail: geoamin@gmail.com.

conditions, with a special emphasis to the changing rainfall patterns and temperature increase, have a primary role in affecting wetlands in Lebanon, notably those with fragile hydrologic balance. This has been also reflected on the major terrain components (i.e., land cover/use) which have been also changed over the last few decades. Hence, the area of these wet landforms is reduced due to the decreased volume of feeding water, and this has been exacerbated in the view of uncontrolled human interference. In this study, two wetlands along the Lebanese coastal zone were investigated using different tools and approaches for analysis. Even though, both selected wetlands are extending along the coast, yet they have different hydrologic regime, and thus they have been affected at different levels. This study introduces atypical example accounting for a comparative analysis between wetlands in Lebanon with different hydrology and aquifer characteristics.

Keywords: groundwater, hydrologic regime, saturated soil, migratory birds, Lebanon

INTRODUCTION

General Overview

There are many definitions for "Wetlands," but the common understanding that they represent water bodies with defined hydrologic regime, which almost reveals water on terrain surface, where unique ecosystems exit. They are also identified as landforms that saturated with water, in which the latter is the dominant factor determining the nature of soil development and the types of plant and animal communities (Cowardin et al., 1992).

Lebanon encompasses a number of localities saturated with water, and makes it a region with distinguished terrain features, landscape and ecosystem. Therefore, these typical landforms are viewed mainly from the frequent existence of water bodies on terrain surface and from the saturated substratum. There are many physical elements that contribute to the existence of wetlands.

Wetlands the Lebanon are found in regions with defined geomorphology and water feeding mechanism, which are governed mainly by the geological and hydrological setting where they occur. They are mostly characterized by low-lands and depressions, long raining periods (November-April) and the existence of aquifers.

In Lebanon, wetlands are utilized for different purposes, mainly for tourism, agronomical production and water supply. In addition, wetlands are known as localities for unique species of flora and fauna. They also represent substantial sites where migratory birds from Africa congregate en route to Europe (e.g., *Pernis apivorus*). Therefore, these water-saturated landforms should be well reserved as they constitute an integral part of the ecosystem in the region.

However, Lebanese wetlands exist with fragile ecosystems and they are extremely threatened by urban expansion, which have discouraged the continual development of wild life. However, their preservation is important not only for ecological and economic reasons, but also because they constitute a precious treasure of the country's natural heritage (El-Khouri, 2012).

The coastal wetlands of Lebanon have utmost significance, since they represent remarkable natural sites along the extended Lebanese coastal zone, but they are vulnerable to risk since the coastal zone of Lebanon occupies more than 70% of the country's population, and thus the uncontrolled activities are still exaggerated there.

The obtained studies on wetlands in Lebanon are still rare if compared with their natural importance, and in many cases they only applied to specific themes (Shaban, 2013, Hassan, 2015), or sometimes they are accounted in regional projects (Ministry of Environment-MoE, 2006). A comprehensive knowledge on these wetlands is still lacking. Recently, water resources in these wetlands have been substantially affected by the changing climatic conditions (Shaban, 2014), with a special emphasis to the changing rainfall patterns and increased temperature. Therefore, the major terrain components of the land cover have been also changed over the last few decades. This impact is well pronounced on water resources and it resulted in an obvious loss in the essential components of wetlands' natural value. Hence, the area of these wet landforms is reduced due to the decreased volume of feeding water, and this has been exacerbated in the view of uncontrolled human interference.

Wetlands of Lebanon

The designation of wetlands in Lebanon is not well determined yet and much identification for saturated regions has been illustrated. However, two major dimensions are always considered to presume the existence of a wetland. These are the duration of water saturation (i.e., water volume consideration), and the spatial distribution of the saturated land (i.e., dimension). Nevertheless, there are tremendous localities where water remains for considerable time duration, but with relatively small areal extent. This is well known in low-land regions and depressions where humid climate is dominant, such as in the several geographic patches in mountainous areas of Lebanon.

According to Shaban (2013), there are seven identified wetlands in Lebanon. They comprise three coastal ones (i.e., Palm Islands, Deir El-Nouriyeh-Cliffs of Ras Ech-Chekkaa and Tyre Beach) and four inner ones (i.e., Ayoun Orghosh, Aamiq, Chamsine/Anjar and Aiha) as shown in Table 1.1 and Figure 1.1.

They have a total area of about 16 km^2. These wetlands are characterized by different geomorphology, hydrology, hydrogeology and ecosystem, and they can be attributed to one or more of the following wetland aspects: shallow lakes and ponds, estuaries and coastal wetlands, and some of them can be attributed to more than one of these aspects at the same time (e.g., Tyre Beach).

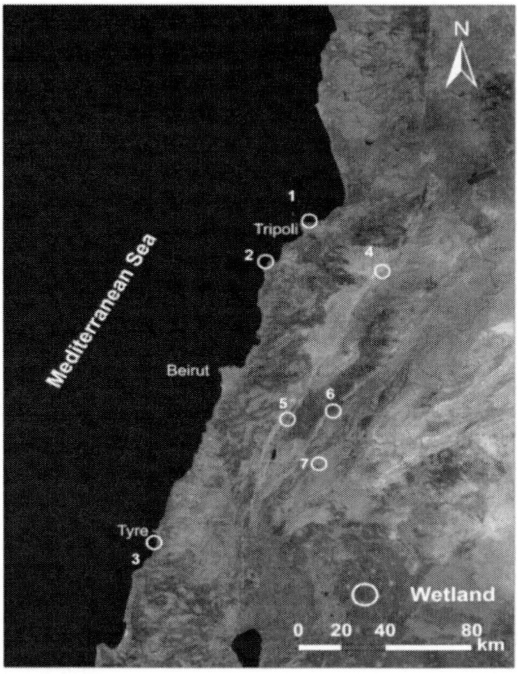

Shaban, 2013.

Figure 1.1. Identified wetlands in Lebanon.

Table 1.1. Identified wetlands in Lebanon*

No.	Name	Area (*km²*)	Description
1	Palm islands	4.1	Three islands almost covered by saline water, notably among the karstic ponds.
2	Cliffs of Ras Ech-Chakkaa	0.85	Jointed cliffs (~200 m) of carbonate rocks adjacent to the sea.
3	Tyre Coast	3.8	Elongated watercourse occurs from a number of artesian springs running to the coast.
4	Ayoun Orghosh	0.47	Surface water in a depression land. They seep from the adjacent mountains, which fed mainly from snowmelt.
5	Aamiq	2.8	Several natural ponds of freshwater in the carbonate rocks.
6	Chamsine/Anjar	0.85	Watercourses running water from Chamssine and Anjar spring and feed Litani River.
7	Aaiha	3.2	Low-lands where water seeps from the neighboring mountain chains.

Shaban, 2013.

Only four wetlands were designated in the RAMSAR List (Ramsar, 2014). These are the: Palm Islands, Deir El-Nouriyeh-Cliffs of Ras Ech-Chekkaa, Tyre Beach and Aamiq wetlands where the last three ones are also considered as natural reserves.

Yet, there is a debate on Deir El-Nouriyeh-Cliffs of Ras Ech-Chekkaa wetland, which was included in the RAMSAR list. Nevertheless, it was reported in that the investigated region is not consistent with the criteria of wetlands. In addition, the exact location of the presumed this wetland was not well determined in the RAMSAR List since 1994. Therefore, two sites were proposed to be investigated in this study, as they were also studied by CNRS and UNESCO-IHP, Paris, in the view one components of Med-Partnership project. These sites are: 1) Cliffs of Ras Ech-Chekkaa along the northern coast and 2) Tyre Beach along the southern coast.

Objectives

The goal of this chapter is to investigate the two selected wetlands and their existing water sources, whether on surface water or groundwater sources, and the related land cover/use components in the coastal wetlands of Lebanon should be: *investigated to assess the impact of climate change and the related human interference on these resources.* This concept was based on field observations and preliminary measures.

Consequently, new management approaches will be proposed to mitigate the physical and anthropogenic challenges to control the sustainability of their natural resources, notably water, in the two identified coastal wetlands of Lebanon.

Therefore, the specific objectives of obtaining this chapter on the two selected coastal wetlands (Cliffs of Ras Ech-Chekkaa and Tyre Beach) can be summarized as follows:

1) Determining the fundamental physical and anthropogenic characteristics of the selected wetlands. This can be achieved by obtaining field surveys, desk studies, data (and information) manipulation and analysis.
2) Identifying the major aspects of water resources (i.e., on-surface and sub-surface), their feeding mechanism and flow regime.
3) Investigating climatic and hydrologic trends over the past few decades and deduce the impact of fluctuated rainfall and increased temperature on the major habitats of the wetlands, including wet zones, water volume, degraded lands, the existing ecosystem, etc.
4) Applying proposed criteria, including the principal components required to categorize the studied site as wetlands.
5) Proposing first-hand elements of the most convenient legislations to be applied in order to conserve the wetlands in Lebanon.

Study Areas

This chapter presents the major physical elements of the selected sites.

Geographic Setting

The two selected sites are coastal localities, according to the scope of the Med-Partnership project. Out of the three coastal Lebanese wetlands, these two sites were chosen because they are terrestrial land forms, fresh-water dominant and in-direct contact with the marine system, and thus the study groundwater can fulfill the need of the Med-Partnership project.

RAMSAR site.

Figure 2.1. Location of Ras Ech-Chekkaa.

Ras Ech-Chekkaa Cliffs Site

This site is located along the northern coast of Lebanon, about 20 km from Tripoli, 45 km from Beirut and 50 km from the northern Lebanese border. It is situated between the following geographic coordinates (Figure 2.1):

- Latitudes: 34° 17′ 38″ N and 34° 18′ 43″ N
- Longitudes: 35° 40′ 10″ E and 35° 40′ 48″ E

The area of the site is about 0.85 km^2, and it represents a plain that surrounded by mountain hills from all sides except the western one where the Mediterranean Sea occurs. It is shaped as an elongated triangle where is base-limb of about 600 m in the southern side and terminates in north with few meters (Figure 2.1).

Tyre Beach Site

Yet, there is a contradictory in defining the exact limits of this site. This is due to the boundary interference between the Natural Reserve and the adjacent refugee camp, as well as the wetland. However; from the hydrological point of view, the wetland and the saturated zone are including the Natural Reserve and the refugee camp too.

RAMSAR site.

Figure 2.2. Location of Tyre Beach Natural Reserve.

This site is located along the southern coast of Lebanon, about 3 km from Tyre, 70 km from Beirut and less than 15 from the southern Lebanese border. It is situated between the following geographic coordinates (Figure 2.2):

- Latitudes: 33° 11′ ″ N and 33° 14′ 59″ N
- Longitudes: 35° 12′ 13″ E and 35° 13′ 27″ E

The site has an area of about 3.8 km^2, represents a coastal plain where the gently-sloping mountains exist at a distance (i.e., few kilometers) to the east and the Mediterranean limits its western side. The site is almost a rectangular shape with about 4 km length and 800 m width (Figure 2.2).

Climate

The two selected sites encompass the Mediterranean climate, which is humid in the wet season to semi-humid in the dry season. Thus, the Lebanese mountain chains, with the highest mountain peaks in the Middle East region (3088 m), sever as a climatic barrier that capture cold air masses from west and condense them as rain and snow (Shaban, 2011).

The wet season coincides with winter period, which lasts from October till March. In winter, the atmospheric pressure perturbations cause abundant rainfall at the coast and on the mountains parallel to it. Besides, the dry season coincides with the summer period, which lasts till September. During this period, no rain is recorded and a state of high pressure dominates the whole country. A typical average transect crossing the country consists of a subtropical coastal climate, a semi-arid continental in the Bekaa Valley and a typical Continental Mediterranean (Chaigneau et al., 2008) climate in the Northern range of Anti-Lebanon Mountains.

More than 90% of the precipitation exists between October and April, and the average rainfall in Lebanon is about 910 mm. This varies between the coastal and mountainous regions. Thus, the estimated precipitation rate on the mountains is about 1350 mm, besides 650-700 mm along the coast. While, it ranges between 650 and 850 mm in Ras Ech-Chekkaa site, and between 550 and 650 mm in Tyre Beach site and the surrounding (Figures 2.3 and 2.4).

The mean annual temperature varies on the coast between 19.5°C and 21.5°C. It decreases approximately 3°C for each 500 m elevation. At 1000 m altitude, the mean annual temperature is around 15°C and decreases to 9°C at 2000 m. Often, the lowest temperatures are recorded in January, the highest temperatures in August, where maximum daily temperatures exceed 35°C in the Bekaa Valley.

Climatic change is well pronounced in the studies sites, similar to the entire climatic of Lebanon. Therefore, there is a remarkable decrease in the rainfall rate which was estimated between 35-40 mm according to the analyzed climatic records of the last five decades. Shaban (2011) considered this decline in rainfall as negligible since it represents only about 4% of the average rainfall rate. However, there is obvious change in the precipitation patterns where torrential rain often exists. In addition, the seasons shifting has been reported (Shaban, 2014).

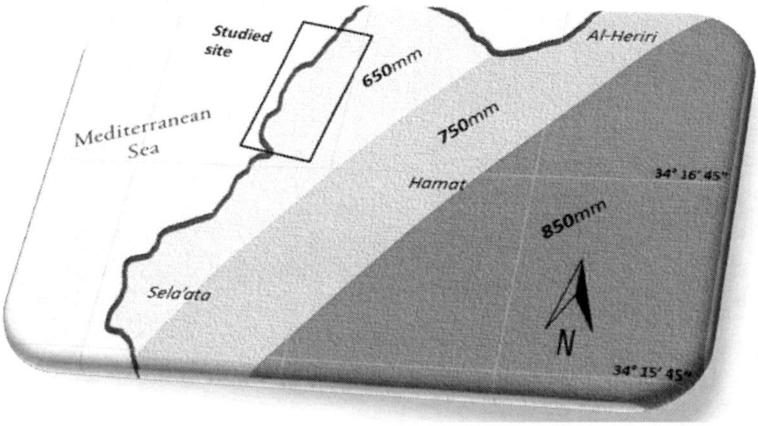

Figure 2.3. Rainfall distribution over Ras Ech-Chekkaa site and the surrounding.

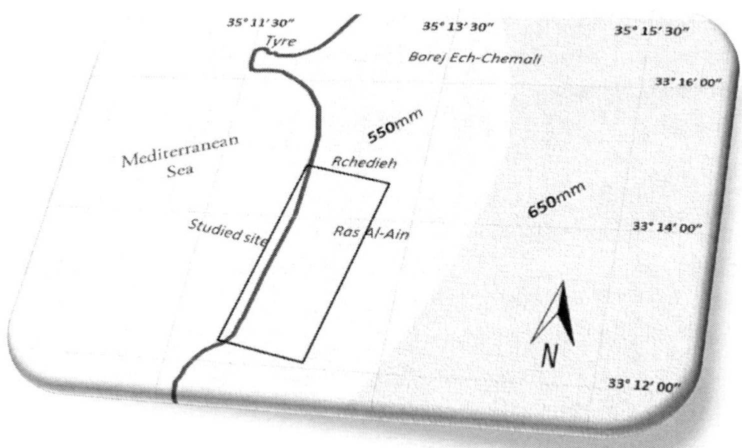

Figure 2.4. Rainfall distribution over Tyre Beach site and the surrounding.

There is remarkable increase in the temperature rate as it was estimated at 1.9°C over the last five decades. This has been well reflected on agriculture, hydrology and it created newly natural phenomena (e.g., erosion, desertification, etc.).

Recently, investigating the aridity index for long-term rainfall and temperature data series indicated that the Lebanon's coastal zone (including the coastal plain and the adjacent mountains chains) is still characterized by humid to semi-humid climatic conditions (Shaban and Houhou, 2015).

Geomorphology

The two selected wetlands are located along the Lebanese coastal stretch, one (Ras Ech-Chekkaa Cliffs) in almost along the northern coast (~50 km from the northern border), and the other (Tyre Beach) at the southern coast (~ 20 km from the southern border). The geomorphological attributes of the selected wetlands are as follows:

Geomorphology of Ras Ech-Chekkaa Cliffs

Even though, Ras Ech-Chekkaa site is located in the flat coastal plain, yet it is geomorphologically is related to the adjacent mountains. Thus, it comprises two major topographic units. These are the mountainous unit to the east, and the confined coastal plain (narrow ribbon of less than 700 m), where the presumed wetland is located, to the west and along the shoreline. The mountainous unit is composed of a miscellany of hills composing a mountain ridge, with average altitudes of about 160 m.

The interface between the two units is represented by sharp cliffs of carbonate rocks. These rocks are always subjected to failure and collapse due to the intensive fractures and joints. In addition the eroded materials form the mountainous ridges are derived along slopes towards the plain and thus hydraulic erosion is a common physical phenomenon.

Three streams occur from the mountain ridge (i.e., at Hamat and Deir El-Nouriyeh villages) and outlet at the cliffs. One stream limits the site exactly from the north; one is approximately in the middle and the third (Wadi Es-Souri) limits the site from the south.

Geomorphology of Tyre Beach

Geomorphologically, Tyre Beach site is also composed of two major morphological units, the mountainous and the adjacent coastal plain, but it is has totally different geomorphology if compared with the previous one. This is because the mountains, which compose a vast plateau region, are far away from the plain which is gradationally slopping from the plateau. In addition, the coastal plain, where the wetland is located, is also a wide coastal ribbon with more than 2.5 km.

The average altitude of the plateau is not well defined, but it can be of about 200 m, while the plain is averaging about 20-25 m.

There are a number of streams span from the plateau towards the plain and they transport tremendous alluvial and fluvial deposits into/and adjacent to the coastal plain.

Geology

The exposed stratigraphic succession in Lebanon, starting from Middle Jurassic, exhibits sedimentation in the marine environment until the Middle Eocene, with carbonate rocks building up the largest part of the stratigraphic column separated by continental sands and other clastics at the base of the Cretaceous and some intercalated volcanics up to the Pliocene. The exposed succession totals a thickness of about 5650 m (Shaban, 2003).

The geology of the studies sites cannot be described separately as the most dominant are the coastal plains, because the existing sediments into these plains are derived from the surrounded elevated regions which have different geological setting. In addition, the surface and groundwater resources are completely controlled by the surrounding rock formations. Therefore, the geology of the selected sites and the surroundings can be described as follows:

Geology of Ras Ech-Chekkaa Cliffs

There are three major geological rock formations constitute the stratigraphic sequence of Ras Ech-Chekkaa Cliffs. They include the coastal plain (where the RAMSAR site exists) extending upward through the adjacent cliffs to include the entire region of Hamat and Dier En-Nourieh villages (Figure 2.5). These geologic rock formations are (in chronological order, from younger to older):

1. Quaternary deposits (*Q*): This comprises less than 15 m thickness sequence (Figure 2.6). It includes a mixture of unconsolidated sediments with alluvial and colluviual deposits where red, sandy soil

is dominant as the major aspect, plus some sandy terraces. Movable dunes also exist at different terrain patches.

2. Miocene Marly Limestone *(m_2)*: It is less than 150 m (Figure 2.6). It is almost massive, moderately hard, thick-bedded yellowish-beige marly limestone with tremendous local fissure and joint system. It constitutes a mountainous ridge extends from Daher Al-Mahhatta from south to Dier En-Nourieh in the north.
3. Senonian Marl (*C_6*): This sequence is characterized by variable thickness from 80 to 100 m in the study area and it is composed from friable, unconsolidated and movable marl with beige color (Figure 2.6). Thus, the friability and softness of these rocks make the region as a bad-land with little uniformity. This rock formation totally shapes the geographic extent of the Miocene formation from east.

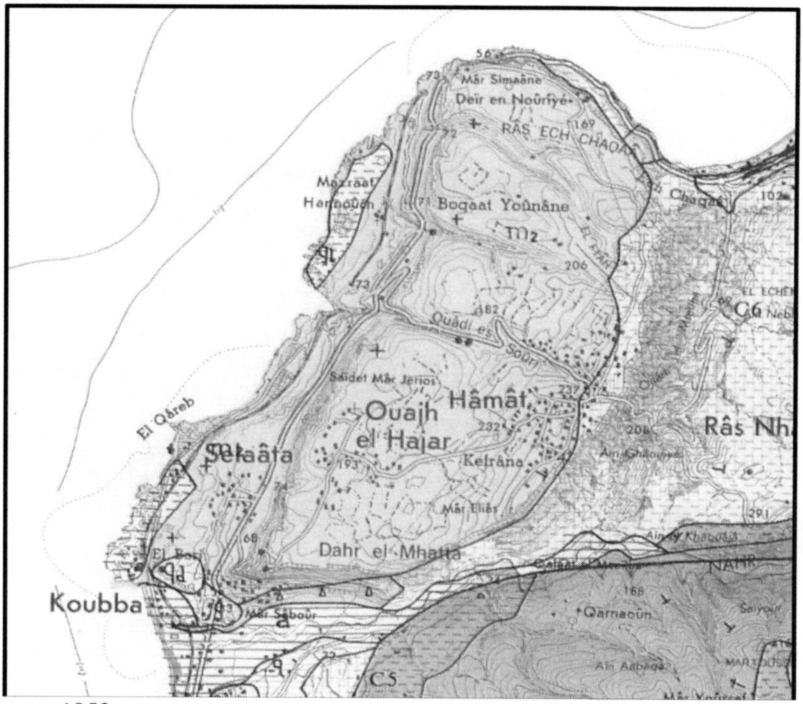

Dubertret, 1953.

Figure 2.5. Geological map of Ras Ech-Chekkaa and the surrounding.

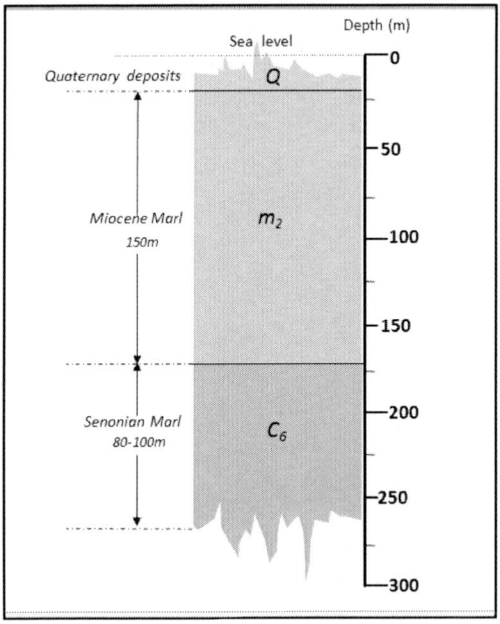

Figure 2.6. Stratigraphic sequence of Ras Ech-Chekkaa site.

Cliffs occur around the massive ridge. These cliffs can be attributed to fault alignment in the eastern side, facing the RAMSAR site over Mazra'at Hanoush area, whilst the cliffs at the northern part of the massive ridge are almost attributed to rotational landslides which constitute movable terrain. This fault, which is of the normal type, has a vertical displacement of about 100-120 m, and it extends in the NS directions and constitutes the cliff themselves of Ras Ech-Chekkaa.

Fissure systems are also dominant in the area of study. These are well developed in the massive marly limestone, notably when the carbonate content increases. Moreover, joints of 2-sets type are well existed at different depths and with a general NS orientation.

Geology of Tyre Beach

The geographic extent of the geological rock formations belong to the investigated site includes also the surrounding regions, notably that water resources exist in the RAMSAR site is totally governed by the remote rock stratum. Even though, this extent is not well defined; however, it can be attributed to the heist mountain ridges from the east, along Kafer Sir, Sir El-Gharbieh, Shehour, Arzon villages (Figure 2.7).

There are four main geological rock formations belonging the studied site is. These are, in chronological order, from younger to older, as follows:

1. Quaternary deposits (*Q*): This sequence has variable thickness, but less than 25 m. It includes unconsolidated sediments mainly of alluvial deposits. Silty soil and sandy terraces are the most dominant in this sequence (Figure 2.8). Movable dunes also exist at different patches.
2. Eocene Limestone (e_2): It relates to the Middle Eocene age where chalky limestone is the principal rock component (Figure 2.8). It is also known by tremendous fracture systems and joints over the whole sequence which is about 180 m in he studied area and the surrounding.
3. Senonian Marl (C_6): This sequence is characterized by variable thickness from 150 to 160 m (Figure 2.8). The major lithology is well bedded, moderately thick marl and marly limestone with almost beige to yellowish-white colour. It comprises dense fissure systems and vertical joints at several localities.
4. Cenomanian Limestone ($C_{4\text{-}5}$): This upper rock formation comprises of more than 130-150 m in the area of interest (Figure 2.8). It composes mainly of massive, thick-bedded, karstified and fractured limestone and dolomitic limestone, which are of grey, yellowish-white color. This formation encompasses a miscellany of fault systems and fractures with different scales.

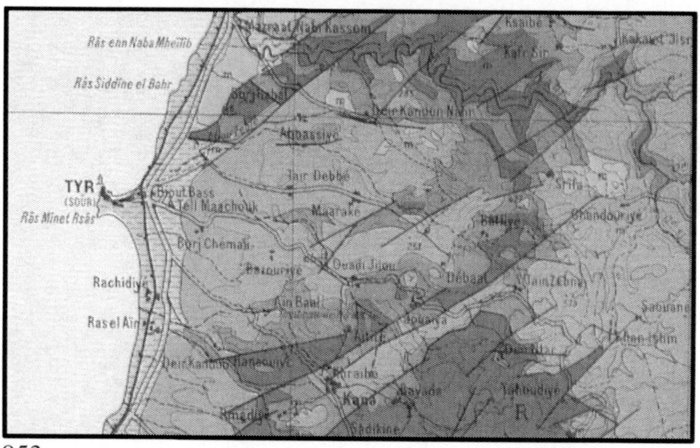

Dubertret, 1953.

Figure 2.7. Geological map of Tyre Beach site and the surrounding.

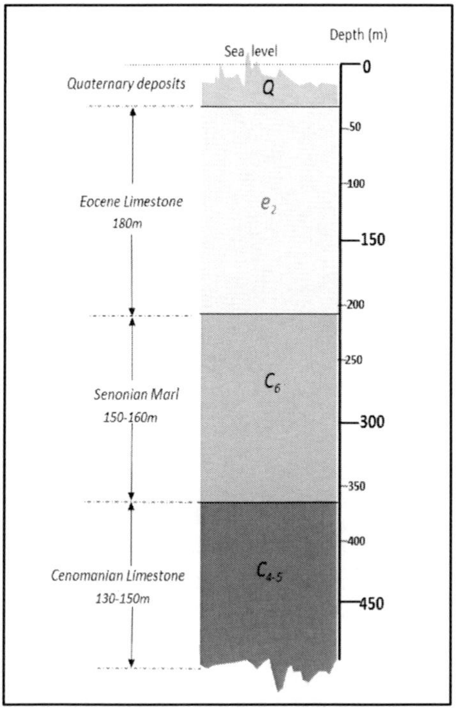

Figure 2.8. Stratigraphic sequence of Tyre Beach site.

Large-scale (i.e., several tens of kilometers) fault systems exist and span from the mountainous regions from the east towards the sea. Most of the existing fault systems in the area are strike-slip type where lateral displacement is dominant (Figure 2.7). Fissures are also dominant deformation aspect where they are well developed in the hard rocks of dolomite and dolomitic limestone. In addition, joints are well observed in many localities, they are almost of multiple-sets where they are clearly occurred in the marly limestone beds.

TOOLS FOR ANALYSIS

This study includes different themes of analysis, thus to accomplish the proposed tasks in this study; it is necessary to utilize a miscellany of tools for investigation and data analysis. This principally includes:

Climatic Data

Climatic data is usually accounted by the records obtained on rainfall and temperature as the major two meteorological parameters that usually measured by the available ground stations. Unfortunately, the institutes who measurethe climatic data in Lebanon have incomplete data records and for limited geographic areas.

Thus, two stations were considered, these are:

1. Tripoli station, which is about 20 kilometer from the investigated site, but along the same longitude. The used datasets are between 1984 and 2013 for rainfall and between 1981 and 2013 for the temperature (i.e., maximum and minimum).
2. Tyre station, which is only few kilometers for the investigated site. The analyzed datasets are between 1985 and 2013 for rainfall and between 1988 and 2013 for the temperature (i.e., maximum and minimum).

Different sources provided the records for rainfall and temperature for these two stations.

This was as follows:

- Lebanon's Atlas of Climate - Atlas Climatique du Liban (CAL, 1971, 1973, 1980)
- General Directorate of Civil Aviation- Direction Générale de l'Aviation Civile (DGAC, 1999)
- Lebanese Agriculture Research Institute (LARI, 2003) with data on Bekaa region from 1954 to1974 and from 1997 to 2003.
- NOAA climatic data, available until the year 1971 (NOAA, 2012).
- Miscellaneous of data from different works and on different geographic locations.

In case of rainfall data missing, there was a support from the remote sensing products. Thus, rainfall data from 1998 up to date was dependent on the remotely sensed data extracted from TRMM (Tropical Rainfall Mapping Mission). TRMM is a joint mission between NASA and the Japan Aerospace Exploration Agency (JAXA). This data is acquired globally.

Hydrologic Data

Surface water is usually included as hydrological data and it implies the discharge of stream flow in rivers and springs. However, in this study the hydrological data is only for the issuing springs in Tyre Beach site, while Ras Ech-Chekkaa site has no defined water source.

Nevertheless, Tyre Beach site has three springs, two major ponds to collect surface water from dug wells, and conveying channels for irrigation purposes as shown in Figure (3.1).

There are flow mters fixed on these water sources (Figure 3.2), which are under the supervision of the Litani River Authority. These flow meters are fixed to measure the stream flow along three localities:

- The Northern Channel (daily data between 1998 and 2014).
- Leakage Channel to the sea (daily data between 1998 and 2014).
- Sea mouth at the outlet (daily data between 1998 and 2014).

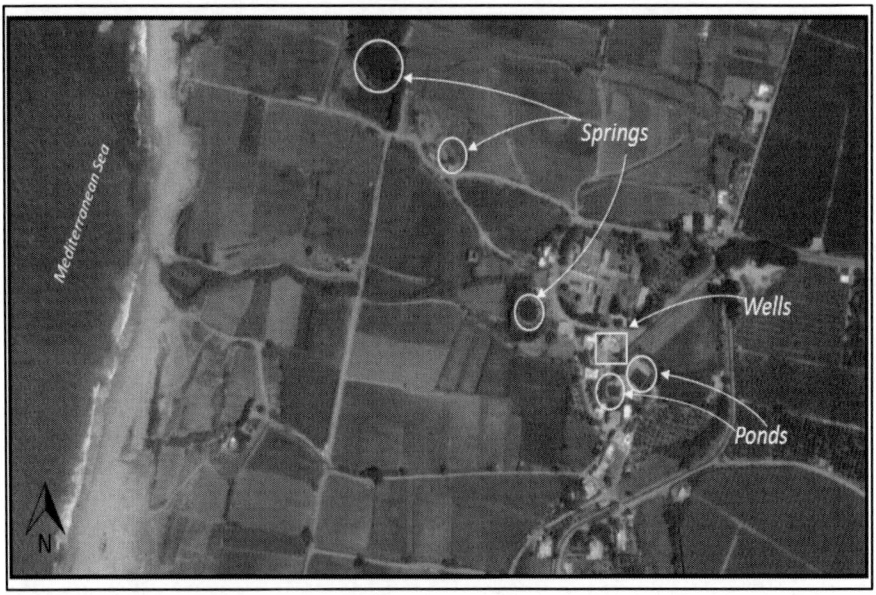

Figure 3.1. IKONOS satellite image (0.83 m resolution) showing Tyre Beach Wetland.

Figure 3.2. Flow meter fixed along the leakage channel in Tyre Beach wetland.

Satellite Images and Softwares

As a useful tool, satellite images were analyzed for different purposes and at different scales. These are digital images with different spatial and temporal resolution. They enable observing all terrain components and calculating their areas.

This includes:

- Geo-Eye-1 images, which were selected for the two sites, with spatial resolution of 1.65 m and for the date of 2014. This type of satellite images, with 11 km swath width, was used mainly for the Land cover maps classification as one of the objectives in this study.
- Aster images with 14 spectral bands as follows: Bands 1-3 VNIR (Visible NIR) 15 m resolution; Bands 4-9 (SWIR) 30 m resolution; Bands 10-14 (TIR) 90 m resolution. The ASTER instrument has a repeat time of 16 days and swath width of 63 × 63 km, with two types of level-1 data: level-1a and level-1b. Level-1a data are formally defined as reconstructed, unprocessed data and consist of the image.

Satellite images were processed using ERDAS-Imagine 9.3 (Leica product) software. It is known as a friendly-user technical tool and has several advantages to identifying terrain features and to calculate and classify each feature observed on the satellite image. While, Arc-GIS-10.2 (ESRI product) was the software of Geographic Information System (GIS) used to manipulate and illustrate the extracted data and information from the satellite images.

Thematic Maps

A number of thematic maps were used in this study. Each map is indicative to specific theme describing defined terrain components, which is needed to be directly/or indirectly analyzed for the two selected sites. The selection of these maps based on the scope of analysis as well as on the availability of these geo-spatial documents. The used maps in this study are:

- Topographic maps, scale 1:50.000 (DAG, 1962)
- Geologic maps, scale 1:50.000 (Dubertret, 1953)
- Geomorphological maps, scale: 1:30.000 (Sanlaville, 1977)
- Land cover maps for the year 2005 (CNRS).
- Soil maps (Darwich et al., 2006)

Field Survey

Several aspects of field survey which were carried out to accomplish this study. They were attributed to different themes of analysis. In addition, the tools used in these surveys were also different and with diverse scale of analysis. Therefore, the field surveys obtained in current study include:

1. Geological and hydrological investigation to identify water sources and saturated zones. We used: GPS, Laser meter, Manometer, geological and geomorphological maps, geological kits, Questionnaire.
2. Land cover mapping to identify the exiting terrain features and to verify the features which were extracted from the satellite images. The used tools are: land cover maps extracted from satellite images, GPS, Spectro-radiometer for selected spots and Questionnaire.

3. Soil investigation to identify soil types and their characteristics. In this survey; however, GPS, soil kits, shawl, Auger and soil maps were used.
4. Species investigation to identify different species per different geographic patch in each site. Hence, GPS, field format, and a Questionnaire were used in this survey.
5. Monitoring migratory birds to identify their number and behavior in the selected sites; therefore, GPS, counter, and Questionnaire was used.
6. Socio-economic survey on water exploitation and different aspects of use and water demands. The Questionnaire was used to accomplish this survey.

METHODS FOR ANALYSIS

There are several approaches of analysis applied in this study, since it includes diverse objectives in characterizing the two selected sites with a special concern to the impact of the changing climatic condition on water resources and the related themes.

Climatic Data Analysis

Gaps were found in the climatic data (i.e., rainfall and temperature) for the two sites. Thus, it was necessary to have complete climatic records in order to deduce the trend in the rainfall and temperature over the last few decades in the areas of interest. Consequently, when the climatic records for rainfall and temperature become ready; therefore, statistical illustrations can be applied to assess their trends.

For the selected sites, the most convenient and available data was as follows:

- Ras Ech-Chekkaa site:
  - Rainfall = 30 years
  - Temperature = 33 years
- Tyre beach site:
  - Rainfall = 29 years

– Temperature = 26 years

The existed gaps in the records were treated by using the "Time Series Forecasting Method" which represents an advanced statistical approach used to interpolate the lacked data among a long-term time series. This method depends on calculating the Moving Average (M_{At}) values for the available recoded years (Y_t). Hence, in this study, M_{At} was calculated using the following formula as an example for the rainfall in Ras Ech-Chekkaa site:

$$M_{At} = (Y_{t-15} + \ldots\ldots + Y_{t-3} + Y_{t-2} + Y_{t-1} + Y_t + Y_{t+1} + Y_{t+2} + Y_{t+3} \ldots\ldots + Y_{t+15})/30$$

When the Moving Average value has been calculated, therefore, the Central Moving Average (CM_{At}) according to the following formula:

$$CM_{At} = \sum_{i=t}^{t+5} M_{At}/6$$

The irregular variations and seasonal variations were calculated by dividing available readings (Y_t) on the Central Moving Average (CM_{At}). In order to calculate the seasonal variations (S_t); however, the irregular and seasonal variations should be calculated among each quarterly. Therefore, the general trend can be assessed from the simple linear regression method by interloping the seasonal variation on Y axis and time variable *t* on X axis. Hence, two coefficients would be resulted. These are: ά for the intersection variable and ᾰ̋ for *t* variable. The two coefficients can be substituted as in the following formula:

$$T_t = ά + ᾰ̋ \times t$$

Hence, the predication for each time interval can be calculated as: $Ft = S_t \times T_t$

Therefore, a complete and comprehensive data time series was resulted for the rainfall and temperature of Ras Ech-Chekkaa and Tyre Beach sites, and this time series data become ready for statistical illustrations to be obtained on Excel format. This will enable to in order identify the general trend in these two climatic parameters, as well as to deduce the ascending/or descending behavior over the investigated time period.

Hydrologic Data Analysis

Analysis of the hydrological data was applied only to Tyre Beach site since Ras Ech-Chekkaa site does not have obvious surface water sources as mentioned previously. Therefore, the hydrological l data analysis included the obtained data from the three fixed flow meters in Tyre Beach site. These are the: Northern channel, Leakage channel and at the sea mouth. Therefore, daily date series was simply converted into monthly in order to facilitate treating the records, as well as to enable figuring out the trending behavior clearly.

Gaps in the available records were treated in the same approach which was used for the climatic datasets was also applied here. Thus, the "Time Series Forecasting Method" and similar statistical tasks were obtained to have a complete time series of data.

In addition to the numerical analysis for the hydrological data, water estimates was identified by inquiring the inhabitants who own the agricultural lands in both sites (Tyre Beach and Ras Ech-Chekkaa). This implies general identification from framers about the behavior and water volume estimates of surface water in each site, since these farmers have been witnessing water regime since several decades ago. This was integrated among the dedicated Questionnaire and it broadly includes the following:

- Sources of water and different consuming aspects.
- If the framers follow specific approaches for measurements.
- How the farmers (and inhabitants) can judge the general trend of water discharge.
- What are the alternative water supply sources in case of water shortage.
- If there are any estimates or discussion mentioned in the Media.

Groundwater Investigation

Groundwater investigation has been applied for the two sites, and it based mainly on the recognition of the controlling geological elements, including mainly the stratigraphy and structures. Hence, when these two geological attributes are correlated to groundwater, they will be described as hydro-stratigraphy and hydro-structure; respectively.

For the hydrogeological investigation, it is inaccurate to deal with the site separately. In other words, the major existing geological system should be entirely identified and classified according to different groundwater characterizations. In this study, the following were the approaches applied in this characterization:

Hydro-Stratigraphy

Geological maps (scale 1:50.000) were primarily analyzed, and accompanied with field surveys in order to analyze the hydrogeological setting of each site from the lithological (i.e., rock type) point of view including the stratification of rock beds. This includes:

- Identify the rock types (i.e., lithology) and their overlying order.
- Attribute each lithology to the water-bearing property. This implies mainly the identification of "*aquiferous*" rock formations.
- Identifying the controlling geologic system which governs water flow and accumulation.
- Identifying the correlation between the issuing surface water sources (i.e., springs) and the hydro-stratigraphic setting.
- The identified hydro-stratigraphy was tightly integrated with the hydro-structure.

Hydro-Structure

In investigating groundwater regime, the hydro-structure analysis is of great importance to, and in many times, it can be more significant than the hydro-stratigraphy, notably in a region like Lebanon where the tectonic activity is well pronounced. Therefore, hydro-structure includes the analysis of rock deformations (e.g., faults, fractures, joints, folding, etc.).

In this study, the geological maps (scale 1:50.000) were of significant use. In addition, satellite images were also utilized. This includes Aster satellite images, which can provide information on the fracture systems at different scales and with diverse orientations.

Aster satellite images, with 14 spectral bands, were processed using *ERDAS-Imagine 9.3.* This in turn required applying a sub-setting procedure to select the needed scene where the optimal sites exist. In order to accomplish this, all images had to be primarily converted to *TIFF* format.

Thermal bands in Aster images were useful to identify wet longitudinal horizons (i.e., indicators for fault alignments). In this regard, Bands 10-14 (TIR) with 90 m resolution were utilized. Consequently, and for better discrimination, a number of optical and digital advantages were carried out in the *ERDAS-Imagine* software. This includes, band combination, enhancing, filtering, color slicing, edge detection, etc.

The combination of geologic maps and satellite images analysis with field verification could help identifying the major hydrogeological controls.

Moreover, the identification of the hydro-structure elements was implemented using several field geological tools, such as the Incline-meter, Range-meter, *HCl* (10% diluted), etc.

The majority of hydro-structure analysis implies the following:

- Identifying the major controlling fault systems (i.e., geographic extent, orientation), and their role in groundwater flow/storage.
- Identifying the fracture systems and their significance in water recharge.
- Recognizing the tilting of bedding planes and their turn in groundwater flow along different stratums.

Soil Classification

Soil mapping was obtained by collecting other data from previous maps (geological, topographic, Land cover, etc.). After undertaking reconnaissance field visit and work, each polygon was labeled with the corresponding soil type. An index map created by the Lebanese Army on a scale of 1:20.000 is followed in creating the soil maps. Arc -View and Arc-GIS version 10.1 are used to create the final 1:20.000 soil maps, and the legend color is derived from the FAO soil color chart. The map projection is Lambert Conformal Conic, Datum Clarke 1880.

Soil classification is based on descriptive and measurable soil features like diagnostic horizons and related characteristics. These are directly connected with the soil forming processes. The soil horizon is divided into Epipedon (0-50 cm) and Endopedon (100 cm). The organic matter accumulation, color changes and textural differentiation are among major criteria for the upper soil layers. In general, clay accumulation, mottling and secondary calcium carbonate accumulations serve as criteria for the designation of lower horizons. The current soil maps are presented using the legend of the World Reference Base for Soil Information (WRB, 2014).

The horizon development and differentiation were considered as the first criteria for soil classification, except where lithology plays the major role in soil formation.

Ecosystem Analysis

Ras Ech-Chakkaa and Tyre Beach sites were evaluated by carrying out a detailed field survey to verify fundamentally the following items:

- Observed activities
- Observed pressures on the ecosystem
- Expected trends of the pressure
- Expected changes/impacts in view of climate change

This is to help evaluating the potential response of ecosystems to the expected change in the occurring climate trends. The assessment was supported by available published information on both sites as well as site visits to approximate current status and the expected impacts.

Land Cover Map and Its Classification

Satellite images have been collected from the GeoEye-1 satellite which was successfully launched on September 6, 2008. It collects images at nadir with 0.41-meter panchromatic (black and white) and 1.65-meter multispectral resolution. This satellite has the most advanced technology ever used in a commercial remote sensing system. In attempting to develop classification systems for use with remote sensing techniques that will provide a framework to satisfy the needs of the majority of users, certain guidelines of criteria for evaluation have been established.

Field verification was carried out to test the accuracy of image interpretation at selected sites and to clarify interpretation assumptions. Global Positioning System (GPS) were used to precisely locate and delineate the ground sites investigated.

Arc-Map 10.3 was the used software for digitizing the available satellite images and building the supporting database. Arc Info engine, available at the CNRS, was used for map joining and, later, for map editing.

Hierarchical classification used in the national Land-Cover/Land-Use has been applied. It covers 72 sub-categories and four levels. For instance, the level 1 covers artificial areas, agriculture areas, wooded lands, grasslands, wetlands, unproductive areas, and water bodies.

Ortho-rectification has been made, followed by a mosaicking of the different images over the Lebanese territories. Field visits were undertaken to interpretation keys for image interpretation.

Satellite images processing of extrapolation based on iterative controls. The process consists of marking the boundaries of areas representing single cover units on the false-color images, using interpretation keys, ancillary documentation or aerial photographs to assign nomenclature headings, as well as extrapolating the established delineation and identification to different parts showing similar characteristics.

In addition, visual interpretation at a scale of 1/5000 over the selected sites and 1/10000 over the other areas for the imageries of 2013 has been conducted to delineate the diverse zone areas. The minimum mappable unit (100 m^2 for the reserve areas and 625 m^2 for the other) was then defined based on the following three basic requirements:

- Facilitates the legibility of the printed map or easy digitization from the interpretation manuscripts;

Provides a representation of the essential features of the terrain in terms that serve the thematic objectives of the project.

RESULTS

Results from different analyzed themes were achieved. This implies: climatic trends, water sources (surface and sub-surface) and their flow/accumulation regime, soil characteristics, the existing ecosystem and the changing land cover/use components.

Climatic Trends

Rainfall Trends in Ras Ech-Chekkaa

The applied statistical approaches on rainfall records helped filling data gaps for Ras Ech-Chekkaa region. Thus, results show slight decrease in the overall rate by about 12 mm over the last 30 years (1984-2013) as shown in, which reveals the annual rainfall dataset (Figure 5.1).

The average rainfall in Ras Ech-Chekkaa region, where the RAMSAR site occurs, is about 686 mm. However, the entire trend shows much more oscillations before the year 2004 (Figure 5.1), where some rainfall peaks exist (i.e., 1986, 1991 and 2003) after which a remarkable stability was noticed. Even though, the dates between these peaks were more or less at similar time interval, which could be as cyclonic interval.

Rainfall Trends in Tyre Beach

The analyzed rainfall records, also after obtaining statistical approaches to have continuous record, for Tyre Beach region, show slight decrease in the overall rate by about 38 mm over the last 29 years (1985-2013) as Figure 5.2, reveals the annual variation in rainfall time series.

The average rainfall in Tyre Beach region, where the RAMSAR site exists, is about 648 mm. While, the general trend, which is similarly to that in Ras Ech-Chekkaa, shows oscillations before the year 2003 (Figure 5.2), where a number of rainfall peaks were noticed (i.e., 1986, 1991 and 2001) after which a remarkable stability occurs.

Temperature Trends in Ras Ech-Chekkaa

The temperature datasets were elaborated for both the maximum and minimum records since both can influence the behavior of tis significant metrological parameter and thus result impacts on different water-related sectors. Thus, the overall trend of the average temperature (i.e., maximum and minimum) was illustrated and based on the manipulation approaches on the available records, which were discussed previously.

The maximum temperature in Ras Ech-Chekkaa region is increased by about 1.25°C, whilst the minimum one was also increased by about 0.81°C and the average temperature in this region has been increased by about 1.0°C over the last 33 years as shown in Figure 5.3.

The average monthly temperature was estimated at about 21.03°C for the time interval between 1981 and 2013, which is above the average temperature of the entire Lebanon (i.e., 19.18°C).

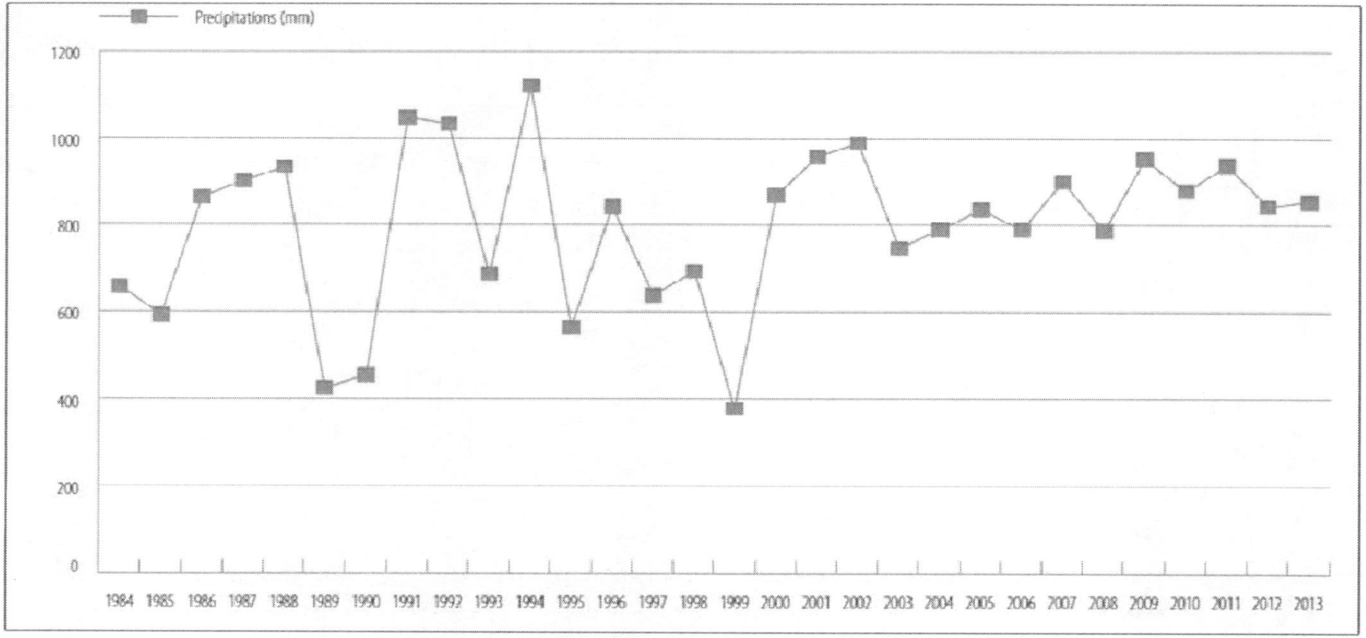

Figure 5.1. Average monthly rainfall trend for Ras Ech-Chekkaa region.

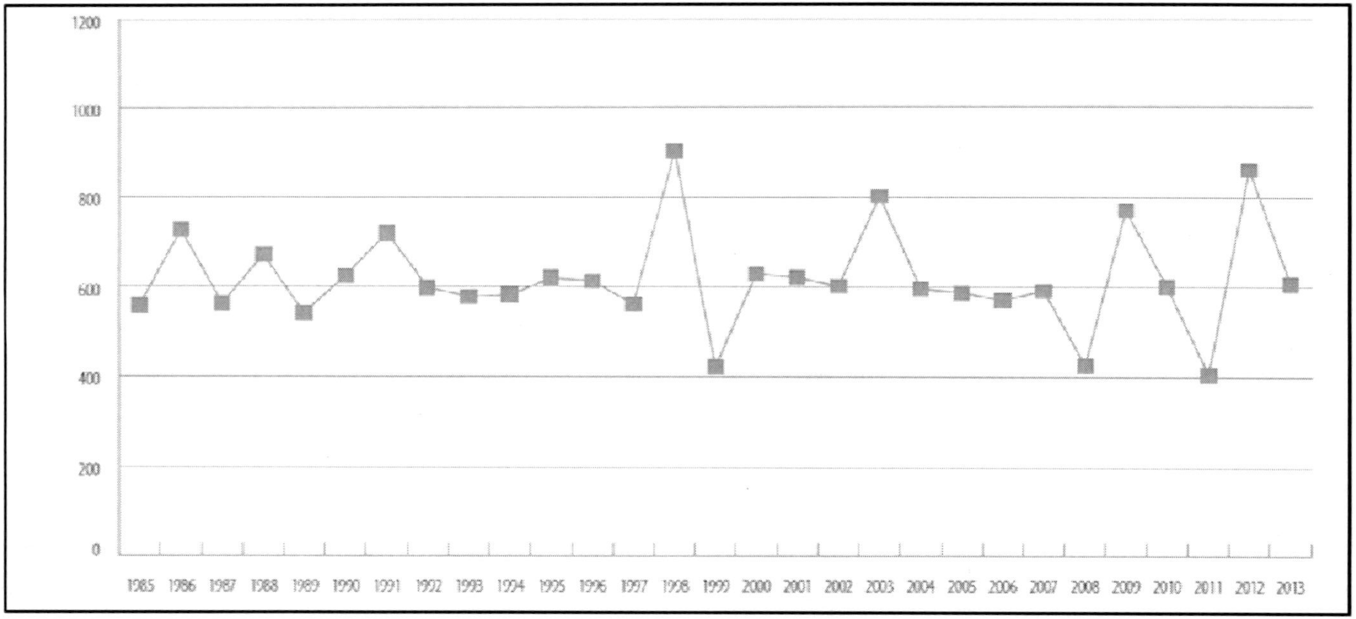

Figure 5.2. Average monthly rainfall trend for Tyre Beach region.

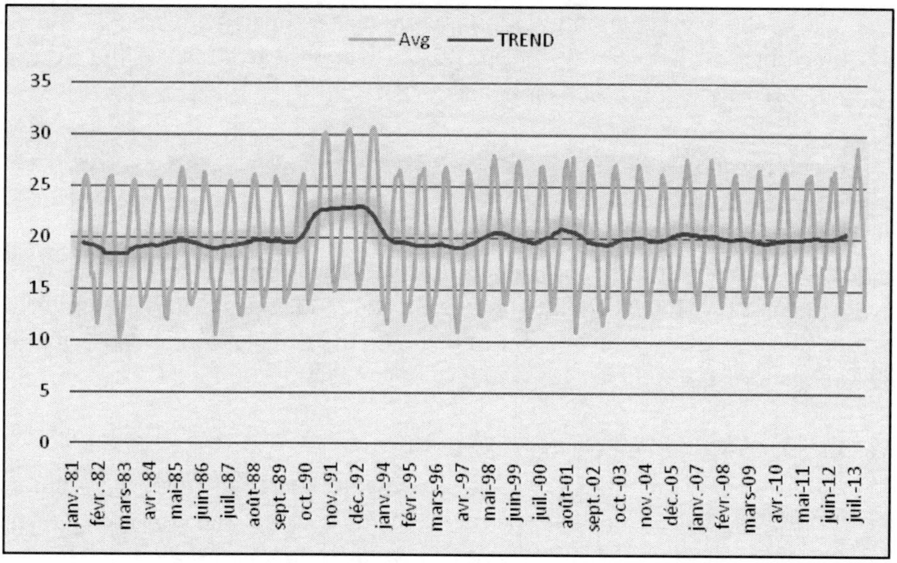

Figure 5.3. Average monthly temperature trend for Ras Ech-Chekkaa region.

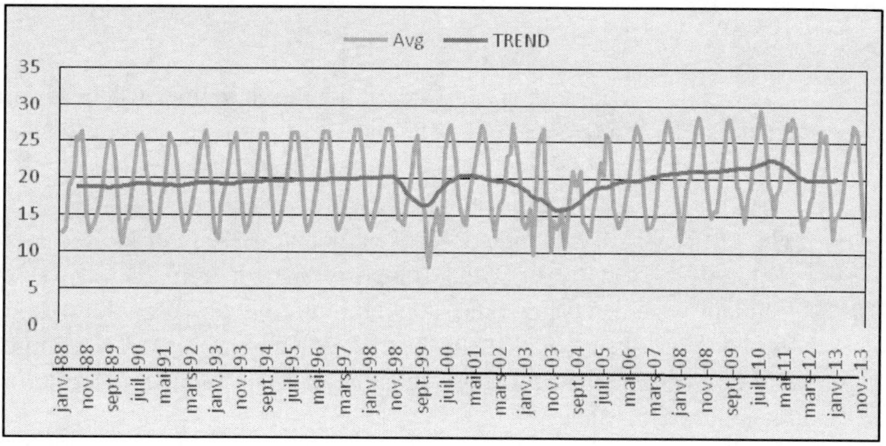

Figure 5.4. Average monthly temperature trend for Tyre Beach region.

Temperature Trends in Tyre Beach

The calculated maximum temperature in Tyre Beach region is increased by about 1.6°C, besides an increase in the minimum values which is about 1.0°C. Thus, the average temperature in Tyre Beach region has been increased by about 1.4°C over the last 26 years as shown in Figure 5.4.

While, the average monthly temperature was estimated at about 21.2°C for the time interval between 1988 and 2013, which also is above the average temperature of the entire Lebanon.

Water Resources

Water is the major component in wetlands, and thus a terrain without any aspect of water cannot be considered as "wetland." Therefore, there is much concern for the volume of water among the wetland zones, and the monitoring approaches are usually carried out to estimate the discharge or water retention in these landforms.

Resources in Ras Ech-Chekkaa Site

As previously mentioned, there is no obvious source of fresh water among Ras Ech-Chekkaa site, which was designated as a wetland according to the RAMSAR List of 2014. This would deserve more attention to have a creditable declaration about this site. Nevertheless, there can be two aspects of water which might be the reason why this site has been accounted as a RAMSAR site. These are:

- There is interaction of the saltwater along the shorelines where several clusters of salines occur. In addition, there are some terrain patches (directly on the coast) which are flooded by tidal water, but they are very limited in terms of areal extent.
- The cliffs, which are composed of carbonate rocks and located at an altitude of about 50 m with a range of 600 m from the sea, show frequent water seepages (i.e., slow and intermittent sheet flow) along the bedding planes of the limestone and dolomite rock. This limited and non-continues water flow feeds the adjacent sediments located in the plain of this site.
- The three streams which are extending from the mountain ridge (i.e., at Hamat and Deir El-Nouriyeh villages) and cut the cliffs towards the coastal plain are also contributing in the saturation of sediments of this plain, but it is also at a limited range.
- There are well known sources of submarine freshwater discharge (i.e., sub-marine springs) into the sea beside Ras Ech-Chekkaa region. This was detected by airborne thermal infrared survey (CNRS, 1999).

Based on these concepts about the designation of the Cliffs of Ras Ech-Chekkaa as a RAMSAR wetland site; however, no considerable source of water to be accounted, and thus no measurable aspect for water volume has been applied in this site. Nevertheless, the obtained Questionnaire with the inhabitant in Mazra'at Hanoush region indicated that the coastal plain (RAMSAR Site) was witnessing saturation in soil for a couple of months, but this has no longer appearance in the past few decades.

Changing Water Volume in Tyre Beach Site

As it was discussed earlier, Trye Beach site has several water sources, whether on surface or subsurface. These are:

- 3 water well
- 3 springs
- 2 ponds (Figure 5.5 5.6)
- 2 major conveying channels (Figure 5.6)
- A number of small artificial tributaries.

However, hydrologic analysis to assess the changing volume in water resources in this site was applied to the datasets of the three flow meters which are fixed at: the Northern Channel (deriving water from wells), at the Leakage Channel to the sea and the at the sea mouth.

Figure 5.5. A major water pond in Tyre Beach site.

Figure 5.6. Water pond with a conveying channel in Tyre Beach site.

The results can be summarized as follows:

1. Northern Channel: It extends about 5 km from the ponds of Ras Al-Ain towards Al-Ma'ashouk in the north. The average discharge from this channel is about 12.4 million m^3 per year. The general trend shows an overall increase in the discharge volume at about 3% over the last 16 years as resulted from the stream flow time series analysis of the available datasets (Figure 5.7).
2. Leakage Channel: It extends less than 1 km from the ponds of Ras Al-Ain towards the sea as an outlet channel to be utilized by farmers for irrigation purposes. However, there is pumping from this channel for irrigation purposes. The average discharge from this channel is about 0.86 million m^3 per year. While, the general trend shows an overall decrease in the discharge volume at about 72% over the last 16 years as observed from the stream flow of the analyzed datasets (Figure 5.8). This abrupt decrease is mainly artificial, and it is attributed to the control applied from the pumping stations of water wells which are managed by the Litani River Authority (LRA), and thus LRA regulating water outlet to the sea for several supply considerations.
3. Discharge at sea mouth: This is a locality where the flow meter is fixed at the sea mouth. Thus, the average discharge from this channel

is about 16.3 million m^3 per year. Generally, the discharge trend is ascending at about 9% over the last 16 years as observed from the stream flow analysis (Figure 5.9).

Hydrogeology

Hydrogeological investigation includes groundwater behavior in terms of water flow and storage among different rock formations. It is the common approach to monitor groundwater mechanism to wards different processes, as well as hydrogeological investigation is a known approach of analysis for groundwater exploration.

Even though, the two studied sites have totally different hydrogeological setting, yet, the hydrogeological investigation in this study is an essential part for the principal objectives, and this includes mainly the relationship between the existing groundwater in the wetlands and their feeding regime. Thus, the changing in the behavior of groundwater governs the wetland characterization and the related disciplines.

Hydrogeology of Ras Ech-Chekkaa Cliffs Region

Hydro-Stratigraphy

Lithology (i.e., rock type) has a major role in controlling water flow from terrain surface into the underlying stratum (i.e., percolation), and then governing the mechanism of groundwater flow between different lithologies until water and accumulates or discharges into a defined outlet, which is usually into the sea.

Hydrogeological investigation of the RAMSAR site (i.e., wetland) includes the entire mountain chains adjacent to the site from east, which is represented by the rock masses of the Miocene marly limestone. The selection of this geographic extent based on the concept that entire groundwater feeding zone must be considered. It is; therefore, constitutes a land area extends from Daher Al-Mhatta from south to Dier En-Nourieh in the north (Figure 2.7).

In addition to top-soil and mixed deposits of the Quaternary age, there are two major lithological types in the region of Ras Ech-Chekkaa Cliffs. These are the Miocene Marly Limestone (m_2) and Senonian Marl (C_6), where the second one (C_6) is not well exposed in the limits of area of investigation (except from the eastern part), but it has a role in the hydrogeological regime, notably that the overlying Senonian marly limestone is almost uncovered from its all sides and has a thickness of less than 150 m only.

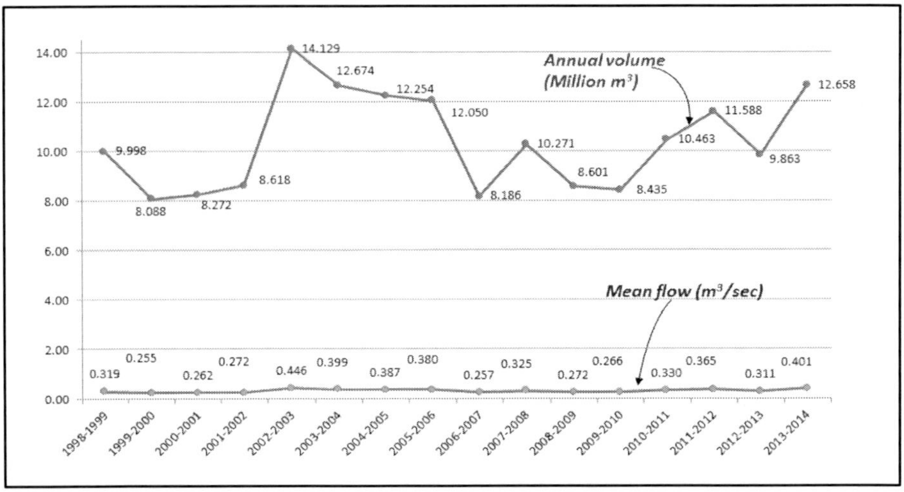

Figure 5.7. The general stream flow in the Northern Channel at Tyre Beach region.

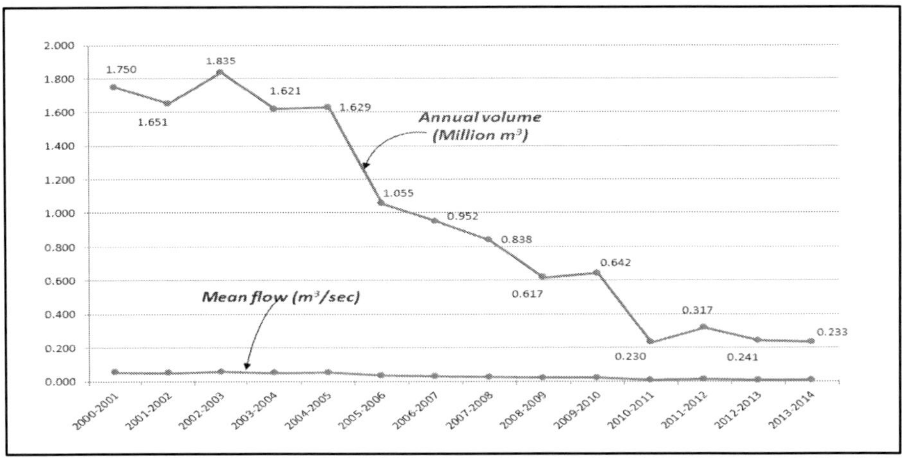

Figure 5.8. The general stream flow in the Leakage Channel at Tyre Beach region.

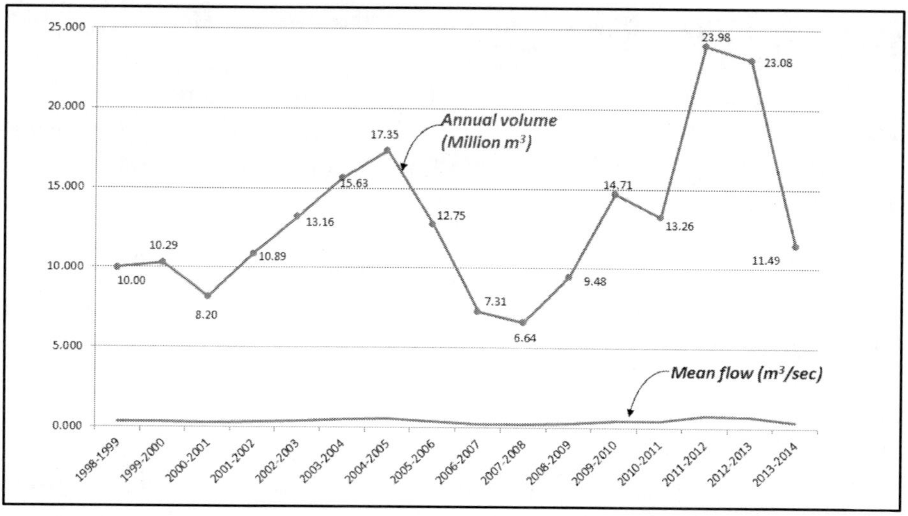

Figure 5.9. The general stream flow at the sea mouth of Tyre Beach region.

Table 5.1. Hydrologic characteristics of the different rock formations in Ras Ech-Chekka site and the surrounding

Rock formation	Thickness *(m)*	Infiltration* *(%)*	Permeability	Porosity	Aquiferous characteristics
Quaternary deposits *(Q)*	<15 m	<15	Variable	Variable	Non-defined
Miocene marly limestone *(m_2)*	<150 m	22	Very low	High	Aquiclude
Senonian marly limestone *(C_6)*	(exposed) 80-100 m	10	Moderate	Moderate	Aquitard

* Infiltration percentage from the precipitated water.

Hydro-Structures

Deformations in the area of study play a role in controlling groundwater regime (i.e., flow and storage). Thus, the existed fault alignment which represents the boundary between the selected site (i.e., RAMSAR site) and the mountain ridges makes these mountainous area as a boundary-exposed rock masses where water at the western side of theses masses almost fee to flow (i.e., not confined).

In addition, the fractures and local fissuring systems enhance the vertical and lateral flow of groundwater among rocks, while joints have well influence in groundwater flow, which is almost vertical and little lateral flow was accounted, except that between the contacts of different beds.

The dominant dip (i.e., inclination of rock beds) readings show a value range between 8 and 28 degrees with a general dip direction westward. This is well pronounced on the rock masses of the Miocene marly limestone, and specifically at the cliff exposures.

Hydrologic Characteristics

The three existing rock formations in the study area encompass diverse hydrological properties, which are dependent on the lithological characteristics of rocks and then govern groundwater transfer from different rock strata as well as the restriction of movement at different localities (vertically and horizontally). This is represented fundamentally by permeability and porosity of rocks (Table 5.1).

Groundwater Recharge and Flow Mechanism

The mountains adjacent to the RAMSAR site from east are the only catchment (surface and subsurface) source of water; however, the annual estimated water volume captured (V_c) in this area can be calculated as follows:

Captured water volume = Catchment area x Average annual rainfall

$$V_c = A_c \times R_a$$

$$= 26\ (km^2) \times 686\ (mm)$$

$$= 26 \times 10^{-6} \times 686/1000 = 1.78 \text{ million } m^3$$

About 50% of this water volume, which is received from rainfall, is returned back to the atmosphere as evapotranspiration, and thus the rest volume (about $900 \times 10^{-3} m^3$) will be as surface run-off or infiltrates downward in rock stratum. According to the UN (1967) estimates, the infiltration rate of the Miocene marly limestone (m_2) is 22%; therefore, the estimated recharged groundwater is about:

$1.78\ x^{-6} \times 10^{-3} \times 22/100 = 0.391$ million m^3/year

The amount of water enters among the exposed rock of the Miocene marly limestone (m_2) percolates (infiltrates) among its rocks and mainly through the dominant vertical fractures and joints. Also, the infiltration processes is developed between different bedding planes which are well marked in this rock formation.

Since the general inclination of the bedding plans of the exposed Miocene rock formation is westward, thus the general groundwater flow will be to the west as shown in Figure 5.10. This flow will continues vertically and horizontally unless groundwater reaches the Senonian marl formation (C_6), which is a geologic formation with very low permeability, and thus groundwater flows almost horizontally along the contact between the two formations (m_2 and C_6). Therefore, the general flow in the study region is westward (Figure 5.10). Thus, there will be seepage along the bedding planes which appears as slow and sheet flow from the cliff, as well as a higher flow rate exists at the bottom of the cliff at the contact with C_6.

Hydrogeology of Tyre Beach Region

Hydro-Stratigraphy

Studying the hydrogeological setting of the RAMSAR site (i.e., wetland) includes the entire mountain hill located to the east of the site. In terms of hydrogeological boundary, this comprises a relatively vast area with unidentified limits. Thus, the expected areal extent can be to the most elevated mountains, representing the alignment (from north to south) joining the villages of Kfar Sir, Baflay and Deir Ntar (Figure 2.9). This geographic extent of the studied area was built on the concept that entire groundwater feeding zone must be considered.

The Quaternary deposits of mixed alluival sediments are dissipated at different localities of the area, notably in depressions. Hence, three major lithological types exist. These are (from newer to older): Eocene Limestone (e_2), Senonian marl (C_6) and the Cenomanian limestone ($C_{4\text{-}5}$). Nevertheless, the altitude of these overlying rock formations is opposite to their chronology (i.e., newer is al lower altitude and vice versa).

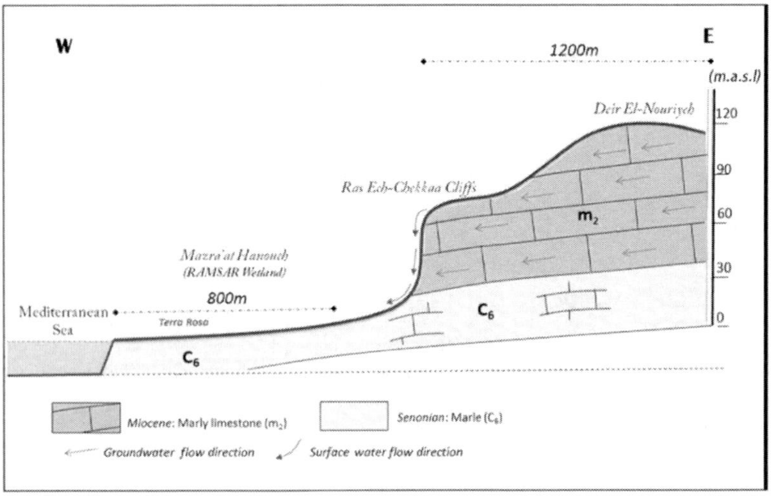

Figure 5.10. Hydrogeological cross-section in the RAMASAR site of Ras Ech-Chekkaa Cliffs and the surrounding area.

The Cenomanian limestone ($C_{4\text{-}5}$) constitutes the principal recharge zone at the top mountains region, and then the Senonian marl (C_6) has very little recharge, and again the recharge becomes more developed in the exposed Eocene Limestone (e_2).

There will be diverse percolation rates from the region adjacent to the coastal plains where the selected RAMSAR site is one of them. However, the existence of colluviual and alluvial and fluvial deposits at the foot-slope of the mountainous region (in contact with plain) makes it as a barrier where groundwater can irregularly outlets.

Hydro-Structures

Geologic structures are well developed in the study, and owing a miscellany of aspects and scales, and hence governing the entire groundwater regime. This is mainly represented by the strike-slip faults which span from the heighted altitudes of the recharge zones, where Cenomanian limestone ($C_{4\text{-}5}$) is well developed, reaching the coastal plain as shown in Figure 5.11.

Even thought, the geological maps obtained by Dubertret, 1953 shows a number of fault system, yet the processing of Aster satellite images, notably the use of thermal bands could identify the geographic extent of many faults in the area as shown in Figure 5.11. The used satellite images enabled recognising the exact faults' extent, notably in the areas where friable and quaternary deposits exist.

Fractures systems are also exist with essential contribution in groundwater regime. This includes a variety of fissures at different scale and orientation. Hence, these fissures and in combination with the existing joints, increase the recharge (vertically and horizontally) property of the exposed rocks. They are well developed in the hard rocks rather than those with marly content.

Karsitfication is another hydro-structure phenomenon which dominantly occurs in the carbonate rocks of the area, and it is increased where the dolomite content is found. Therefore, the karstified rocks show a number of carbonate dissolution aspects. These are mainly as: caves and conduits, where the latter may extend to several hundreds of meters depth with irregular orientations, but they found to be well excavating along the surfaces of weakness (e.g., contact between rock beds, fracture alignments, etc.).

The predominant bedding planes inclination (between 12 and 24 degrees) westward, and thus governs the general groundwater flow seaward. Thus, the dip is found to be increased in the hard carbonates (i.e., C_{4-5}).

Hydrologic Characteristics

As a result of the differentiated lithological characteristics of the exposed rock formations in the study, hence diverse hydrological properties exist, and this results different groundwater regime, notably in the context of the mentioned dominant hydro-structural properties. In other words, the area of study owns the extreme hydrological aspects, especially between the limestone and dolomite (with higher permeability) versus marl (with lower permeability).

Table 5.2. Hydrologic characteristics of the different rock formations in Tyre Beach site and the adjacent mountainous regions

Rock formation	Thickness *(m)*	Infiltration* *(%)*	Permeability	Porosity	Aquiferous characteristics
Quaternary deposits (Q)	<25 m	<15	Variable	Variable	Non-defined
Eocene Limestone (e_2)	180	23	Moderate-high	Moderate	Aquiclude
Senonian marly limestone (C_6)	150-160 m	10	Moderate	Moderate	Aquitard
Cenomanian Limestone (C_{4-5})	(exposed) 130-150 m	35	High	High**	Aquifer

* Infiltration percentage from the precipitated water.

** High is due to secondary porosity as a result of extensive fractures.

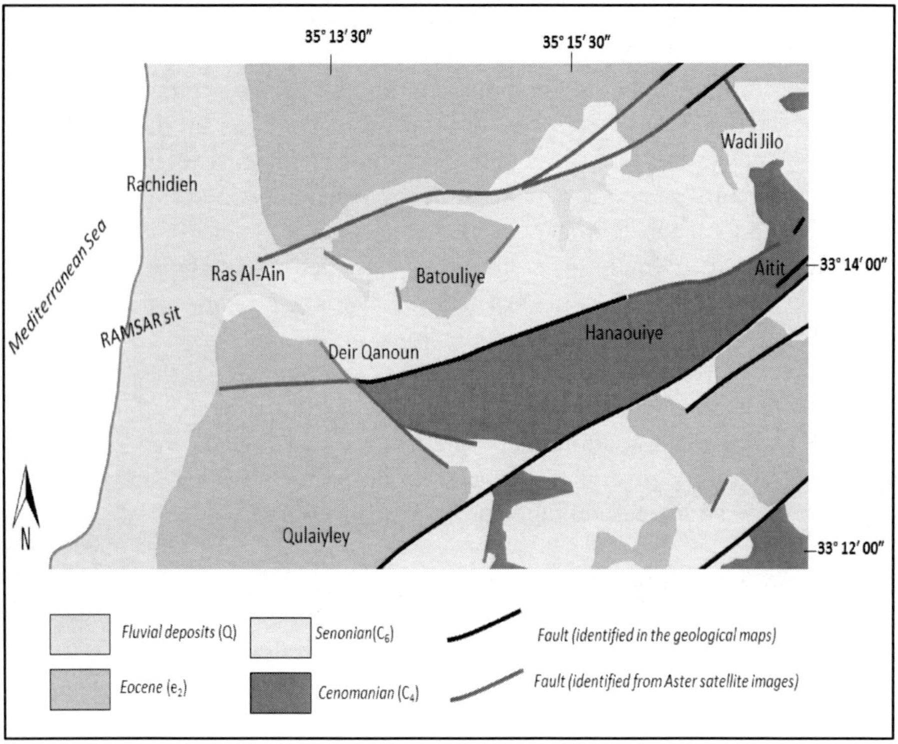

Figure 5.11. Faults identification from the processing of Aster satellite images.

Aquiferous characteristics of the rock formations in the study area can be summarized in Table 5.2.

Groundwater Recharge and Flow Mechanism

Similarly to the concept followed for Ras Ech-Chekkaa site, the mountainous regions adjacent to the RAMSAR site (Tyre Beach wetland) to the east, which is the major recharge (catchment) zone for the surface and subsurface water sources; however, the annual estimated water volume captured (V_c) in this area can be calculated as follows:

Captured water volume = Catchment area x Average annual rainfall

$V_c = A_c \times R_a$

$= 465\ (km^2) \times 648\ (mm)$

$= 465 \times 10^{-6} \times 648/1000 = 301$ million m^3

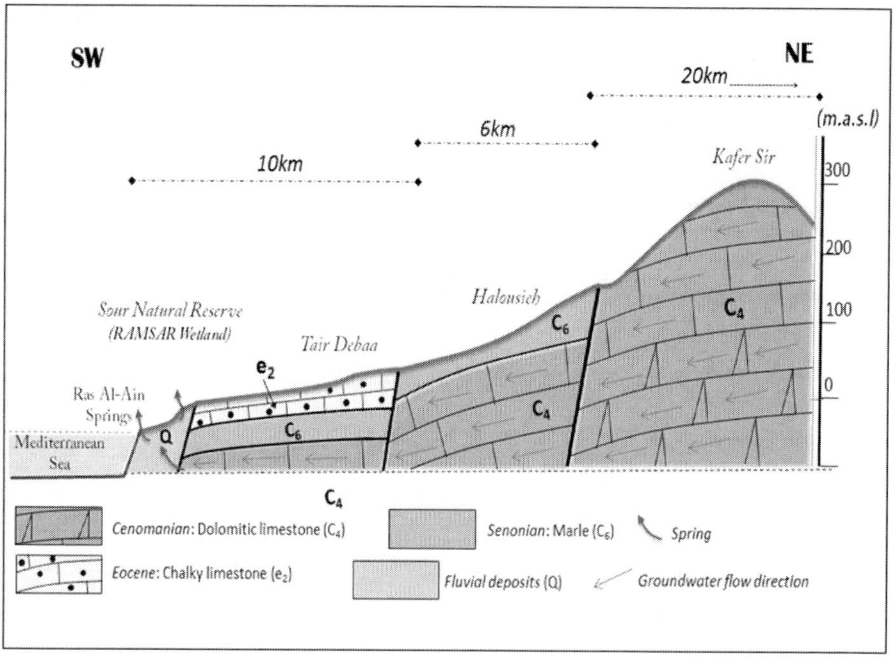

Figure 5.12. Hydrogeological cross-section in the RAMASAR site of Tyre Beach and the surrounding area.

Considering that about 50% of this water volume will return back to the atmosphere as evapotranspiration; therefore, about 150 million m^3 will remains to feed surface water runoff and infiltration. Hence, considering the estimated infiltration rates for each exposed rock formation calculated by UNDP (1970) and the calculated geographic area of each formation; therefore, the average infiltration rate will be about 29%. Consequently, the estimated recharged groundwater in the study area will be about:

$$301 \text{ x } 10^{-6} \text{ x } 29/100 = 87 \text{ x } 10^{-6} \text{ m}^3/\text{year}$$

This volume of water can be considered as an annual renewable source of groundwater that feeds the mountainous region adjacent to the coastal plain where the RAMSAR site of Tyre Beach is located. The majority of this volume is easily flows along the tilting bedding planes, faults systems and karstic conduits towards the sea, i.e., westward (Figure 5.12).

It is well identified from the built stratigraphic cross-sections that this groundwater has a subsurface connectivity between the carbonate rocks of the

Cenomanian formation (C_{4-5}) as shown in Figure 5.12. This connectivity is enhanced by the bed inclination as well as by the existing structures. Nevertheless, the presence of marly beds in the Eocene formation (e_2) and the Senonian marls (C_6) makes the Cenomanian rock formation (C_{4-5}) as a confined aquifer that inject water westward (to the coastal plains) as appear in Figure 5.12. Consequently, the relatively high velocity of groundwater, estimated at 2-5 cm per second (Shaban, 1985), will flow from the carbonate rocks along the inclined bedding planes, as well as groundwater movement along the lateral-displacement faults and the existing karstic conduits will be interrelated and outlets its load at the contact between the coastal plain and rock masses of the mountain hills where a thick sequence of alluvial and mixed sediments occur; therefore, groundwater will be injected at different localities in these sediments resulting the discharging springs of Ras Al-Ain, as well as the groundwater in the dug wells in the same area (Figure 5.12).

Soil Characteristics

The area of Tyr Beach is spread over quaternary deposit over level land south of City Tyr. The soil types are mainly deep and introduced since long times in the agricultural and other human activities. The first impression when visiting this area is the high human pressure and large agricultural activities and excess irrigation caused by two main factors: Water availability year round in the absence of water harvesting goes directly to the sea if not used directly from the source or stream. The prevalent primitive methods of surface irrigation known for low efficiency and large amount of water infiltrated into the groundwater poses additional risk to the groundwater quality as water table can be enrich with soluble nutrients or pollutants. The high level of nutrients found in the soils of the area support the hypothesis of large soil-groundwater contamination with nitrates and possibly other harmful chemicals.

There are three main soil types found in the area of study: From the east to the west, a narrow strip of Hypocalcarric Fluvisols and Haplic Vertisols can be found in association over quaternary sediments (Figure 5.13). They are followed by haplic Fluvisols and Arenosols.

The first type is usually located on undulating microrelief and is better drained. The Vertisols type is found on flat areas and represent deep crack upon drying. Fluvisols are rich in fine tubular pores indicating good soil biological activity. The near beach area formed by sand is also partly used in agriculture under low plastic tunnels which leaves a very narrow strip partly non used by human sand beach that can merely be considered as wild life area for any aquatic animals given the high disturbance and compaction of the ground as a result of vehicles traffic.

Ras Ech-Chekkaa area is located on the sea shore of the North Lebanon and spread on slopping and level land. The soil formation was affected by the slope and lithology formed by hard limestone and quaternary sediments. The soils formed on the slopping lands are shallow and eroded. The soil is intermittent with rock outcrops. In the middle of the area, remaining of old terraces could be observed.

The most widespread soil of the Chekka wetland is the Eutric Leptosols. This most abundant soil, developed on fluvial deposits, is shallow and well structured (Figure 5.14) in a level plain and flat regional slope. The profile is imperfectly drained with slight sheet erosion is observed throughout. The content of rocks and coarse fragments is high with common very fine tubular and interstitial pores.

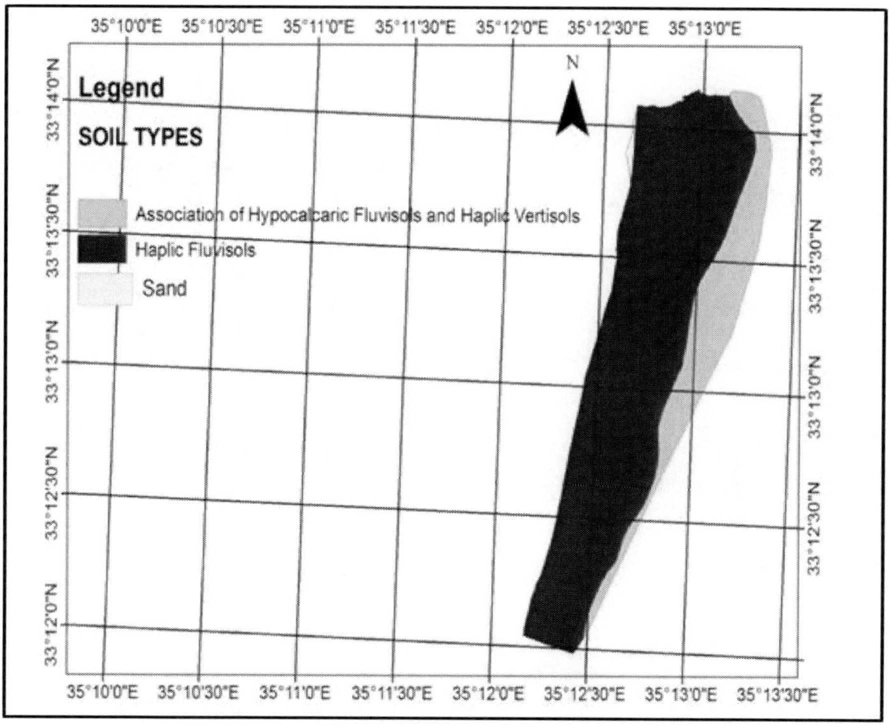

Figure 5.13. Soil map of the Tyr Beach site.

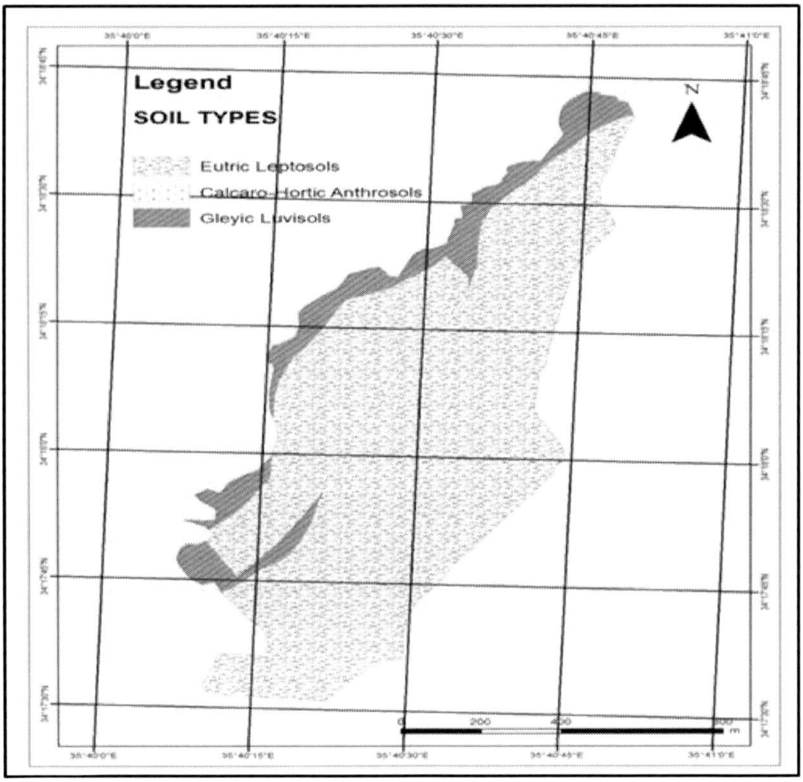

Figure 5.14. Soil map of Ras Ech-Chekka site.

Figure 5.15. Spartium junceum.

The eastern border of the wetland is surrounded by a shallow eroded soil developed in a medium gradient mountain and rolling regional slope on limestone lithology and classified as Lithic Leptosols. Despite the observed signs of sheet erosion and rockiness, the soil is also enriched with fine tubular pores. The Gleyic Luvisols is spread along the seashore. The soil is non saline and non calcareous, deficient in nitrogen and enriched with potassium and phosphorus indicating recent past use in agriculture.

Ecosystem

Ecosystem of Ras Ech-Chekkaa

According to AECID/MOE/UNDP (2011), the site is characterized by hills; an undisturbed maquis paints the steep escarpments of limestone and covers a plateau of woodland and scrubland. Those types of ecosystems form just 6% of the land area in Lebanon, and 6-10% of the Batroun district revealing the significance and rarity value of the investigated site.

Conservation Status and Recognition

The site can be considered as a sensitive and a hotspot area, preserved but threatened, requiring protection (UNDP, 2011), ill-defined with no management or conservation plan. Additionally, the concerned area is located on a significant Middle East migratory birds route and embraces habitats for migrating passerines, which classifies it ultimately as an IBA area (Important Bird Area).

The wooded landscapes are increasingly rare and demand, where the site is located holds natural beauty, biodiversity importance, and historical interest that need preservation in their current status and protection from possible challenges and impacts. In this concern, this site is proposed to be classified as a nature reserve, under the category IV of "Protected area with sustainable use of natural resources" according to the International Union for Conservation of Nature (IUCN).

Ecological Evaluation

Flora Description

Ras Ech-Chekkaa is a habitat for a diverse floral species, mentioning notably Cyclamen and Narcissus species (AECID/MOE/UNDP, 2011). Harboured with a dense evergreen oak, the woodland reaches a height of 3 m, along with an abundant understory of shrubby species as such: *Pistacia lentiscus*, *Calycotome villosa, Spartium junceum* (Figure 5.15), *Sarcopoterium spinosum*, with few scattered carob trees *Ceratonia siliqua* on the lower altitudes. Thus, this zone is a dazzling junction of two vegetation series of Thermophillous *Quercus calliprinos* vegetation community and the Carob-Lentiscus series dominated by the Lentisc. An aquatic species *Arundo donax* is also encountered there as a characteristic plant of wetlands. lists the existing communities of trees, shrubs, climbers and herbs recorded in the studied site (AECID/MOE/UNDP, 2011; EcoMed, 2012).

Fauna Description

Numerous marine species are assembled in this littoral area, being a representative site of marine environment and biogeographic region. Supra-, medio- and infra-littoral zones were described in Ras Ech-Chakkaa marine ecosystem (RAC/SPA-UNEP/MAP, 2015) with submarine freshwater springs that amplifies the marine diversity. Sixteen fish species are recorded, notably some endangered sea turtles are also found in the concerned site: the Green turtle (*Chelonia mydas*) and the Loggerhead turtle (*Caretta caretta*). The site is a sensitive coastal habitat for Vermitid platorms species of highly protection attention, and two commercial sponges: *Hippospongia communis* and *Spongia officinalis adriatica* (AECID/MOE/UNDP, 2011).

Terrestrial fauna is diverse and sometimes unique the site. There are shortened to few Red Foxes *Vulpes*, Feral Dogs, Hyena, Wolf *Canis lupus*, Golden Jackal *Canis aureus*.

Reptile species limited in distribution and variety only allowing mainly *Laudakia stellio* being the dominant species found with *Cyrtopodion kotschyi* among the bare rocks and soils. The Middle-Eastern *Testudo terrestris* is recorded in the scrublands of the region at low densities and two very common species of skinks are also encountered.

Migratory Birds

Five dominant bird species exist in/and around Cliffs of Ras Chekkaa site. The lowering water-levels leading and the intrusion of salt water in the place of fresh water which leads to the salinity of water, has played the major role in the decreasing of the abundance of visiting birds in this area or the decreasing of time spent in this location by the migratory birds in their journeys.

Between 1994 and 2014, there is a decrease in the number of 4 species out of 5 (which is equivalent to 80%) that visiting Ras Ech-Chakkaa in their yearly flyway (Table 5.3). However, the Red-rumped Swallow *Hirundo daurica* seems to have an increasing in its number over this period. This increasing was explained by the conservation efforts that are becoming visible taken by the countries where those species are passing yearly.

Ecosystem of Tyre Beach

This Marine/Coastal wetland is characterized by its ecological, marine and coastal ecosystem values. Five types of habitats are scanned in this site: sand dunes of Mediterranean coast, coastal freshwater land, agriculture land, sandy beaches and pebble and stony beaches. The tremendous site shows hence considerable aesthetic, conservative and recreational values as it draws a mosaic of habitats of the unique coastal dunes in the country and the wide sand beach.

Conservation Status and Recognition

Tyre beach site has been considered as a nature reserve since November, 1998. In 1999, it was declared site 980 in Ramsar Convention. The Reserve is part of the city of Tyre, which was designated in 1984 by UNESCO as a World Heritage Site emphasizing the reserve's archeological importance. Under the convention of SPA, the concerned reserve was classified as Special Protected Area in 2011 and SDTAL considers it part of the littoral regional natural park perimeter.

Internationally, the site has attracted wide interest for holding the following characteristics:

- Important nesting site for globally threatened sea turtles (IUCN Red List).
- Beach of high width and important sand dunes diversity (Riddiford, N. pers. comm., 2003).

Table 5.3. Numbers of bird species between 1994 and 2014 in Cliffs of Ras Ras Chakkaa

Years/ number per species	White Pelican *Pelecanus onocrotalus*		Red-rumped Swallow *Hirundo daurica*		House Martin *Delichon urbica*		Wren *Troglodytes troglodytes*		Blackbird *Turdus merula*	
1994	Stable		Stable		Stable		Stable		Stable	
2014	Decreased	↘	Increased	↗	Decreased	↘	Decreased	↘	Decreased	↘

- Wetland of international significance, designated as a Ramsar site number 980.
- Important resting site for water birds and waders.
- Important remaining coastal agriculture site, where open field crops are planted.

Ecological Evaluation

Flora Description

The Tyre Beach Reserve is classified as "Sand Dunes of Mediterranean Coast" type of the Corine's classification, and belongs to the "White Sand Dunes of the Mediterranean dominated by *Ammophilia arenaria*" habitat (code 16.2122) with *Echinophora spinosa, Eryngium maritimum, Euphorbia paralias, Cutandia maritime, Medicago marina,* and *Anthemis maritima*. The mixture of marine and terrestrial ecosystems made from agriculture lands, coastal sand dunes, wide sandy beaches, marshes and springs found in the reserve is a call of interest of different and diversified floral species assembled along the shift from a type to another (UNEP, 2011).

The site counts 275 species distributed over 50 families as it is noteworthy home to 7 regionally and nationally threatened species, 4 endemic mentioning the Beirut milk-vetch *Astragalus betrytheus* and 10 rare species worth mentioning the *Ficus sycomorus* and *Pancratimum maritimum*, whilst 59 species are restricted to the Eastern Mediterranean area. The concerned site is additionally to note the presence of 50 bio-indicator species among the list of Tyre plants as well as 25 medicinal species (MOE/ UNDP, UL, 2004).

Fauna Description

Thirteen mammal species have been counted in the region and they are distributed over 8 families (Tohme, H. in Jaradi, G. et al., 2004). Two of them, *Meles canescens* as well as the *Acomys dimidiatu* were recorded to be globally threatened and nationally susceptible in Lebanon. However, two pest species were identified in the site, the domestic rat (*Rattus rattus*) and mouse (*Mus musculus*), and are usually indicators of organic waste accumulation. displays the list of mammal species found at the concerned site, reported in 2004 (MOE/UNDP/LU, 2004).

Additionally, 18 herptile species were recorded on previous assessment of the reserve. The three marine turtles (*Chelonia mydas, Caretta caretta caretta, Dermochelys coriacea coriacea*) constitute the only globally threatened species in this reserve. The *Acanthodactylus shreiberi* is endemic and all the others are regionally threatened herptile species (UNEP, 2011). As this site is a stronghold location for nesting turtles, notably the nesting concentrations of the threatened *Caretta caretta* (Figure 5.16).

Migratory Birds

Several species of birds can be observed around Tyre coast including (warblers, waders, stilts, passerines, raptors, kingfisher, gulls, terns and other seabirds). There are 28 dominant species, where 23 of them are of high concern and importance. These can be classified as:

1. 4 globally threatened species,
2. 12 regionally threatened species
3. 7 partially restricted species in the Middle East.

Table 5.4. shows the existing birds species and the changing number between the years 1994 and 2014.

Table 5.4. Numbers of bird species between 1994 and 2014 in Tyre Beach site

Elanus coeruleus Black-winged kite	Phalacrocorax pygmeus, The pygmy cormorant	Pelecanus crispus Dalmatian pelican	Falco naummani Lesser kestrel	Crex crex. Corn crake	Botaurus stellaris Eurasian bittern	Ciconia ciconia, White stork	Aythya nyroca Ferruginous duck	Years/number per species
Stable	Stable	Stable	Stable	Stable	Stable	Stable	Stable	1994
Stable	Stable	Stable	Stable	Stable	Stable	Stable	Stable	2014

Gyps fulvus, Griffon vulture	Accipiter brevipes Levant sparrowhawk	Aquila pomarina Lesser spotted eagle	Falco cherrug Saker falcon	Falco biarmicus Lanner falcon	Gallinago media Great snipe	Larus hemprichii Sooty gull	Glareola nordmani Black-winged pratincole	Years/number per species
Stable	Stable	Stable	Stable	Stable	Stable	Stable	Stable	1994
Stable	Stable	Decreasing	Stable	Stable	Stable	Decreasing	Stable	2014

Pernis apivorus European honey buzzard	Pycnonotus xanthopygos White-spectacled bulbul	Irania gutturalis White-throated robin	Neophron percnopterus Egyptian vulture	Hippolais languida Upcher's warbler	Sylvia mystacea Menetries's warbler	Serinus syriacus Syrian serin	Pernis apivorus European honey buzzard	Years/number per species
Stable	Stable	Stable	Stable	Stable	Stable	Stable	Stable	1994
Stable	Stable	Stable	Stable	Stable	Decreasing	Decreasing	Stable	2014

Land Cover Map

Land Cover of Ras Ech-Chekkaa

For Ras Ech-Chekkaa site, Level-4 land cover map was obtained, which is represented mainly by the flat (i.e., coastal plain) region and the adjacent mountain hills. Thus, the flat area is mainly covered by bare soils and some human activity patches (e.g., urbanism, cultivation, etc.). However, the mountainous area is characterized by bare rocks and shrubs.

Major Terrain Components

According to the processed satellite images (e.g., GeoEye-1 satellite) which were verified in the field surveys, there 8 major terrain components recognized. These are list in Table 5.5, where each component shows a specific area according to the analysis of 2014.

It is clear that the principal component is represented by dense forests which is almost distributed in the flat area (i.e., RAMSAR site), and then shrubs and trees exist. In addition, several components have been totally disappeared as shown in Table 5.5 and Figure 5.17. These are mainly for the agricultural areas and lands.

Figure 5.16. *Caretta caretta* Logger-head Turtle.

Table 5.5. Areas of major land cover components of Ras Ech-Chekkaa site (2014)

Land cover component	Area		% of the total area of the studied site
	(km^2)	(ha)	
Field crops as small field and terraces	0.0434	4.34	5.1
Agriculture units	0	0	0
Material extraction site	0	0	0
Dense forests (Oaks)	0.2557	25.27	30.08
Beach areas	0	0	0
Protected agriculture	0	0	0
Shrubs and trees	0.0644	6.44	7.57
Grass lands of medium density	0.4918	49.18	57.85

Changing Land Cover (2005-2014)

As a monitoring approach, the comparative analysis of the two satellite images with different dates enables identifying the land cover change. In other words, the changing terrain components could be geographically recognized including the area and changing direction as shown in Figure 5.17.

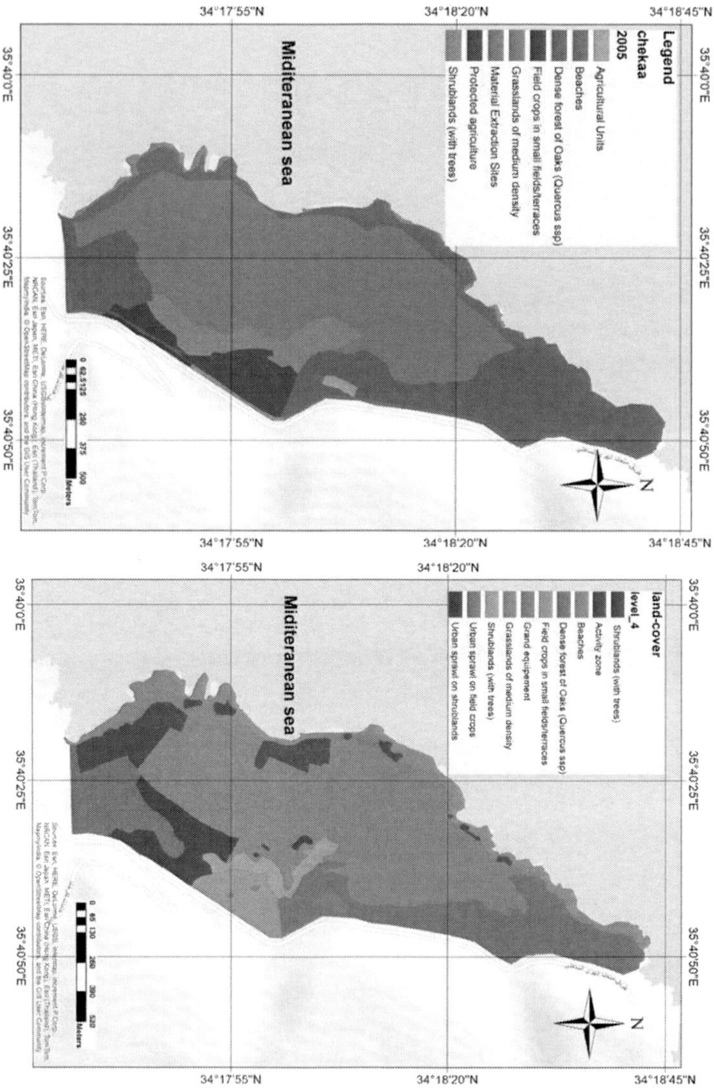

Figure 5.17. Land cover maps of 2005 and 2014 for Ras Ech-Chekkaa site, as recognized from satellite images.

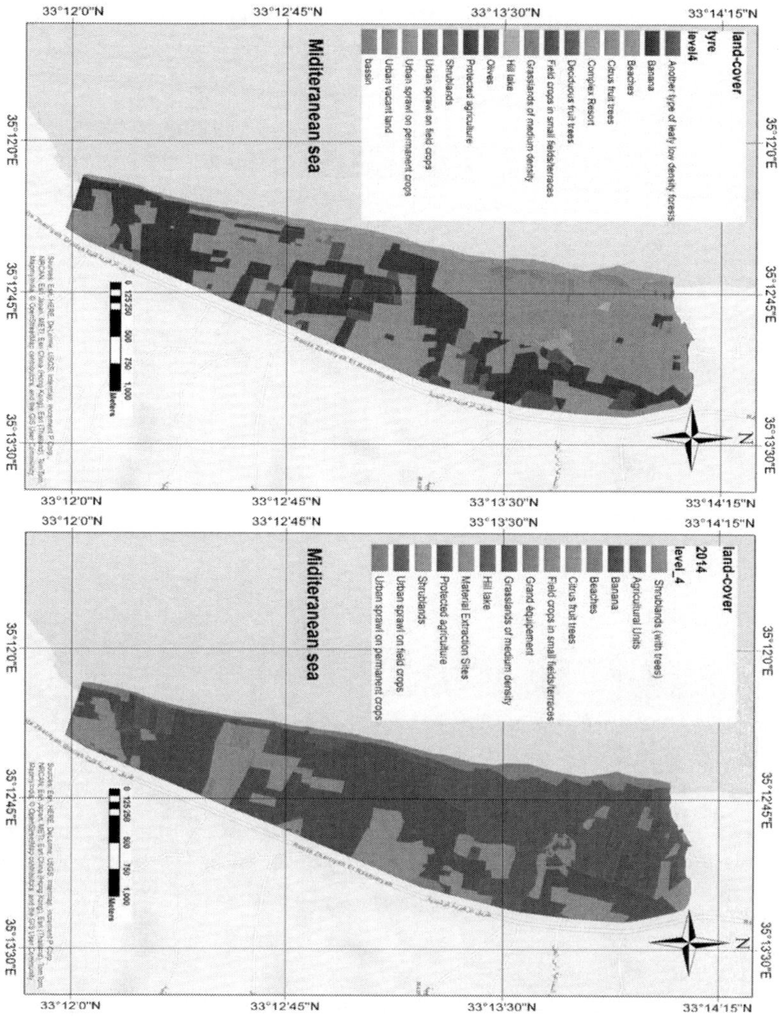

Figure 5.18. Land cover maps of 2005 and 2014 for Tyre Beach site, as recognized from satellite images.

Therefore, the analyzed satellite images for the years 2005 and 2014 enabled calculating the change in area whether increased or decreased for each component. These results were illustrated in Table 5.6. Hence, it was found that there is a remarkable change in the area of sites for extracted materials which was totally disappeared, likewise many other components which were completely evanesced between 2005 and 2014.

Table 5.6. Changing areas of major land cover components of Ras Ech-Chekkaa site (2005-2014)

Land cover component	Area *(km²)*		Changing area *(km²)*	% of changing
	2005	2014		
Field crops as small field and terraces	0.0539	0.0434	-0.0105	19.5
Agriculture units	0.0043	0	-0.0043	100
Material extraction site	2.6960	0	-2.6960	100
Dense forests (Oaks)	0.3126	0.2557	-0.0569	18.20
Beach areas	0.0833	0	-0.0833	100
Protected agriculture	0.0095	0	-0.0095	100
Shrubs and tree	0.0685	0.0644	-0.0041	5.9
Grass lands of medium density	0.5818	0.4918	-0.09	15.47

Land Cover of Tyre Beach

Level-4 Land cover map was produced for Tyre Beach RAMASAR site, which is represented mainly by relatively wide coastal plain adjacent to gradationally sloping terrain to the east. Hence, the majority of the plain is a sandy beach where a number of other terrain components exist, and this includes mainly the cultivated areas of field crops. However, the adjacent slopes also occupy cultivated areas but with fruit trees as a primary type.

Urban settlements and a miscellany of human activities are dissipated in different patches among the site, and the proximity area, but they concentrated mainly along the Tyre-Naquora highway, which represents the contact between the coastal plain and the sloping areas.

Major Terrain Components

The processed GeoEye-1 satellite images were verified in the field surveys to assure the reliability of the obtained results. It was found that there are 18 main terrain components occur in the investigated site as they are list in Table 5.7. Each component shows a specific area according to the analysis of Geo-Eye -1 satellite images of 2014. It was obvious that the majority of terrain components in Tyre Beach site are represented by bananas, citrus trees and file crops.

Table 5.7. Areas of major land cover components of Tyre Beach site (2014)

Land cover component	Area		% of the total area of the studied site
	(km^2)	(ha)	
Citrus trees	0.7114	71.14	18.38
Deciduous fruit trees	0	0	0
Leaf-, low density forests	0	0	0
Banana lands	1.4882	148.82	38.45
harbor basin	0	0	0
Complex resort	0	0	0
Field crops as small field and terraces	1.0013	100.13	25.87
Open fields	0	0	0
Lakes	0.0036	0.36	0093
Urban sprawl on field crops	0.1729	17.29	4.46
Urban sprawl on permanent crops	0.0109	1.09	0.28
Olives	0	0	
Beach sand	0.3851	38.51	9.95
Protected agriculture	0.0540	5.4	1.39
Urban vacant land	0	0	0
Dense urban fabric	0	0	0
Shrub lands	0.0113	1.13	0.29
Grass lands of medium density	0.0397	3.97	1.02

Changing Land Cover (2005-2014)

The applied comparative analysis for the two satellite images of the years 2005 and 2014 enabled identifying the changing land cover over the last ten years. Therefore, the changing terrain components could be identified to presume the areal change and as shown in Figure 5.18.

It is obvious that the main change has been encountered in the citrus lands which have been decreased by about 2.4 km^2, which is equivalent to 77% over the last ten years. Also, there is a decrease in banana land area which represents about 81%. Table 5.8 demonstrates the resulted changes in the area of the major terrain components in Tyre Beach site.

Table 5.8. Changing areas of major land cover components of Tyre Beach site (2005-2014)

Land cover component	Area *(km²)*		Changing area *(km²)*	% of changing
	2005	2014		
Citrus trees	3.1281	0.7114	-2.4167	77.25
Deciduous fruit trees	0.0091	0	-0.0091	100
Leaf-, low density forests	0.0032	0	-0.0032	100
Banana lands	1.8261	1.4882	-0.3379	81.49
harbor basin	0.0010	0	-0.0010	100
Complex resort	0.0319	0	-0.0319	100
Field crops as small field and terraces	0.1650	1.0013	+0.8363	83.52
Open fields	0.0109	0	-0.0109	100
Lakes	0.0053	0.0036	-0.0017	32.07
Urban sprawl on field crops	0.0056	0.1729	+0.1673	96.76
Urban sprawl on permanent crops	0.0462	0.0109	-0.0353	76.40
Olives	0.0039	0	-0.0039	100
Beach sand	0.1924	0.3851	+0.1927	50.03
Protected agriculture	0.0799	0.0540	_0.0259	32.41
Urban vacant land	0.0291	0	-0.0291	100
Dense urban fabric	0.4992	0	-0.4992	100
Shrub lands	0.0121	0.0113	-0.0008	6.61
Grass lands of medium density	0.0420	0.0397	-0.0023	5.47

Conclusion

Wetland Criteria

Wetland, as previously mentioned in the beginning of this chapter, should be characterized by the presence of sufficient amount of water (surface or sub-surface) that makes it a water-saturated zone to attract flora and fauna species different form the surrounding areas.

Therefore, the obtained data and information on the two selected RAMSAR coastal wetlands in Lebanon show typical status for the other existing land forms which are subjected to physical and anthropogenic stresses that make them loss their criteria as a water-saturated terrain, and thus witnessing a decline in the existence of unique flora and fauna, as a result of the regression in water volume.

In this study, three major pillars, with items, were adopted to characterize the existing wetlands in Lebanon. These include mainly: water availability, species uniqueness and state of knowledge. The used criterion, which was applied on the two investigated coastal wetlands, can be summarized as follows:

1. Water availability:
– Existence of visible water sources
• River (in the proximity of the wetland, and water recharge the land for considerable time period as overland flow or as interflow)
• Spring (spring outlets into the surrounding area)
• Natural lake (continuation of water bodies in the proximity of natural lake)
– Existence of groundwater feeding
• Water seepages along rock stratum (groundwater flow along bedding planes)
• Rising peizometric level (continuation of water table above terrain surface)
– Saturated soil
• Level of water retention (well water-situated soil from rainfall)
• Duration of water retention (long-term water saturated soil rainfall)
2. Species uniqueness:
– Unique flora
• Species type (species different from the surrounding regions)
• Species number (species number exceeds the surrounding regions)
– Unique fauna
• Species type (species different from the surrounding regions)
• Species number (species number exceeds the surrounding regions)
3. State of knowledge:
– Reserved area (at the local level)
– RAMSAR site (at regional or international level)
– Natural reserve (at the national level)
– Proposed area (to be reserved and adopted as a reserved site).

However, each component was assigned to criterion priority (i.e., primary, secondary and proposed criteria). In addition, each item in a priority was given a score as follows:

- Water availability = 10 scores,
- Species uniqueness = 5 scores,
- State of knowledge = 2.5 scores.

Moreover, within each item, there are three zones of ranking, where each zone is accounted to 33% of the item score. These zones were described as:

- Well exists
- Normally exists
- Poorly existing

Therefore, the full cumulative score that can numerically characterize each of the investigated RAMSAR sites is 100. Hence, the results were as follows (Figures 6.1 and 6.2):

1. Ras Ech-Chekkaa site = (Primary) 2 poorly existing + (Secondary) 1 poorly existing; 1 normally exists + (Proposed) 1 well exists

= 10 (2 x 0.33) + 5 (1 x 0.33 + 1 x 0.66) + 2.5 (1 x 1) = **14.1/100**

2. Tyre Beach site = (Primary) 2 well exists; 4 normally exists + (Secondary) 4 well exists; 1 normally exists + (Proposed) 3 well exists

= 10 (2 x 1 + 4 x 0.66) + 5 (4 x 1) + 2.5 (3 x 1) = **73.9/100**

Therefore, it is quite obvious that Ras Ech-Chekkaa site does is well consistent with a wetland criterion, since it accounts a low score (~14/100) of wetland criterion. Besides, Tyre Beach site is will demonstrated as a wetland site since is owns considerable criterion score (~74/100).

Physical and Anthropogenic Stresses on Ras Ech-Chekkaa Site

In this study, the existing components of Ras Ech-Chekkaa site were investigated from different aspects to identify its entire setting as a wet landform. The change in these components was obviously noticed from the compared land cover maps. Yet, water bodies do not well pronounced in this site whether on surface or subsurface, and this extends a question on how it was designated as a RAMSAR wetland site according to the RAMSAR wetlands list of 2014?. The answer was demonstrated by several experts and researchers that this sites, including the terrestrial and marine environment, own water saturated patches. This involves: saturated soil, salines, marshes and off-shore freshwater springs. Nevertheless, if this would be considered; however, there are several similar localities along the Lebanese coastal stretch which have similar criterion. In other words, there are many sites along the Lebanon's coast that encompass these components (i.e., saturated soil, salines, marshes and off-shore freshwater springs).

The non-existence of clear water bodies in this site skipped over the idea that it is a wetland, as it was demonstrated by the local inhabitants. Even though, soil saturation does not considerably appear whether in space and time.

The hydrological and hydrogeological investigation shows poor feeding mechanism. Thus, surface water along the three existing streams is limited since these streams are relatively short and have very low water capacity. Besides, groundwater is restricted only to seepages along the cliffs of Miocene carbonate rocks, which outlets sheet water low along these cliffs and presumably feeds the coastal plain where the RAMSAR site is located.

Therefore, the groundwater reservoir has limited extent and capacity (391000 m^3/year) to results continuous feeding to the sites (Figure 6.3). This has been reflected by the negligible decrease in rainfall rate which was estimated at 12 mm over the last 30 years, and this was also influenced by the increased temperature (1.25°C).

It is quite obvious that any changing in the climatic conditions in the investigated site and the surrounding hydrological and hydrogeological system will be critical and abruptly influences water feeding towards the RAMSAR site, and this is the current case. This in turn has been reflected on the presumed wetland components, such as the limited number in flora, and fauna, notably the remarkable decrease in the number of migratory birds, which was estimated at 80% of the known species in the studies site.

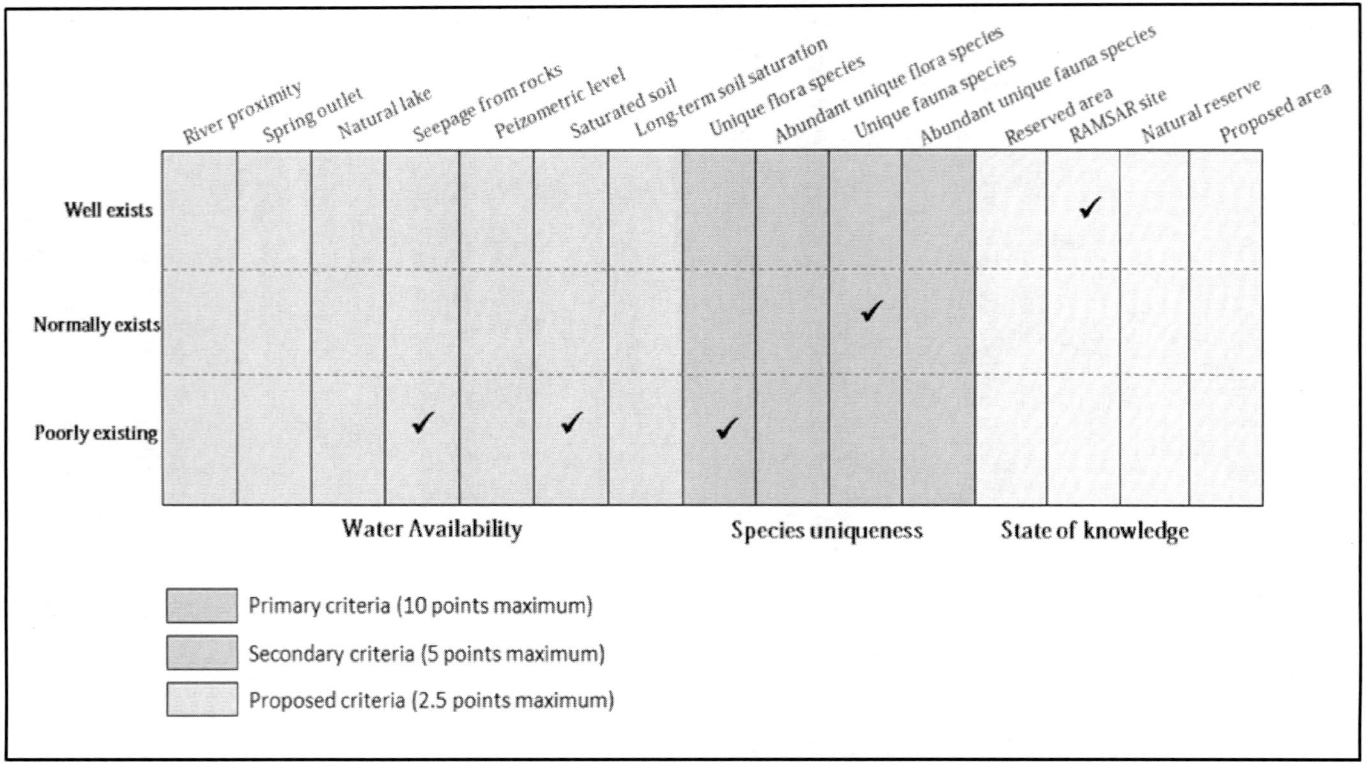

Figure 6.1. Characterization of Ras Ech-Chekkaa wetland site.

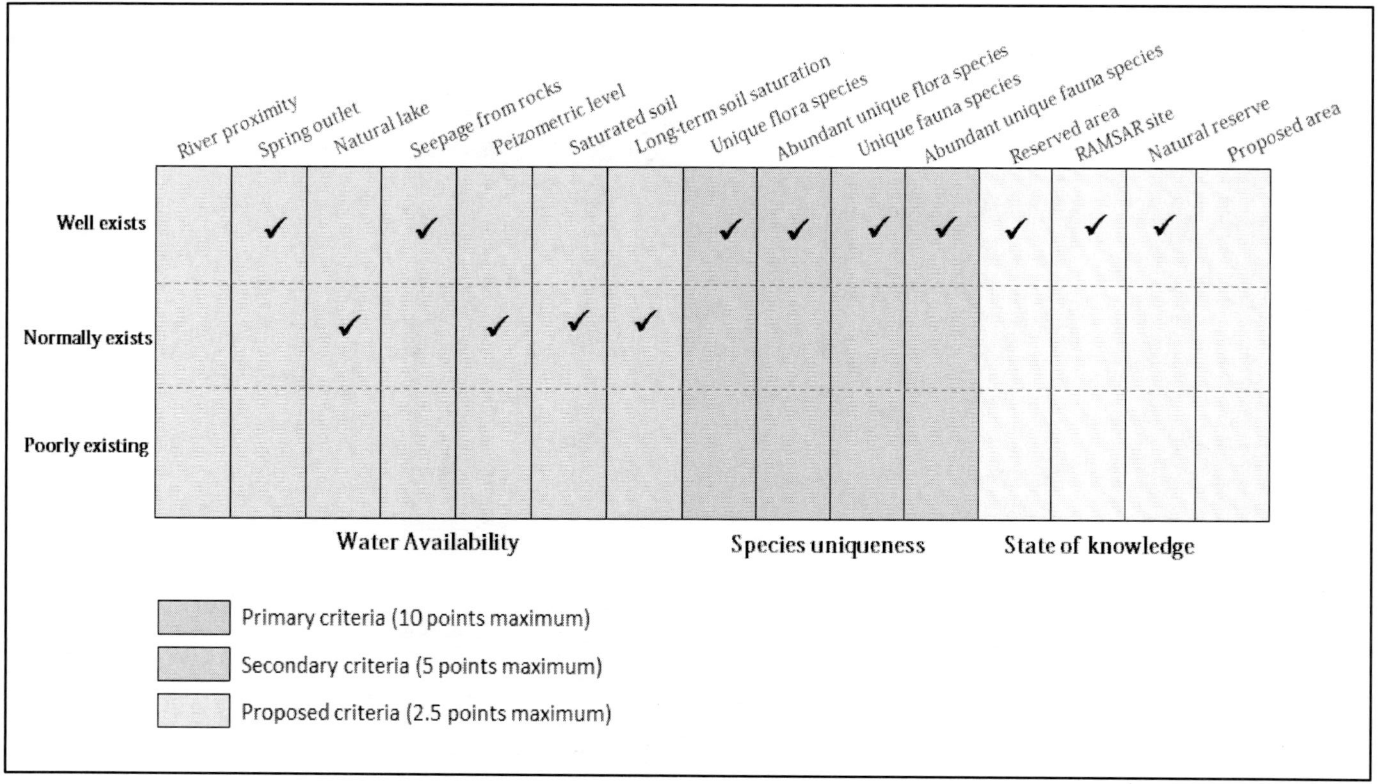

Figure 6.2. Characterization of Tyre Beach wetland site.

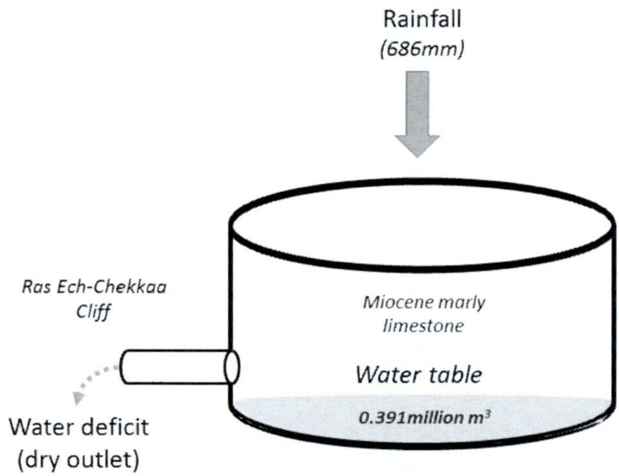

Figure 6.3. Scheme showing groundwater flow mechanism in Ras Ech-Chekkaa site.

Figure 6.4. Overgrazing in Ras Ech-Chekka area.

Human interference is commonly observed in Ras Ech-Chekkaa. This is not limited to those negative activities among the site itself, but extend to the neighboring areas, where water is captured from. These activities include:

- Uncontrolled groundwater pumping
- Wood logging and forest fires
- Over grazing activities (Figure 6.4)

1. Uncontrolled rocks excavation
2. Arable lands fragmentation, mainly by roads.
3. Uncontrolled urban expansion

4. Marine pollution by the contaminated effluents
5. Hunting.

Physical and Anthropogenic Stresses on Tyre Beach Site

Tyre Beach wetland was also investigated and it took more in-depth analysis since the existing components are tremendous, certainly those water bodies, as they were mentioned previously in this chapter.

Even though, there is a clear change in the components of land cover, yet it was a question to be extended, how that the volume of water resources in this RAMSAR site do not show any regression over the last two decades as shown in the results of stream flow analysis?. Even more, the inhabitants declared that no remarkable regression in water volume has been noticed since few decades ago. This is well pronounced and proved since water sources in this site are fed from large-scale groundwater reservoir (i.e., aquifer of the Cenomanian limestone rocks).

Figure 6.5 show a simplified scheme for groundwater feeding from remote areas, but within the same hydrogeological system, where the large-scale faults are extending and transporting groundwater to the site. Even though, the existing streams, along relatively gently sloping terrain, can contribute in flourishing the adjacent coastal plain with water, where the RAMSAR site is located.

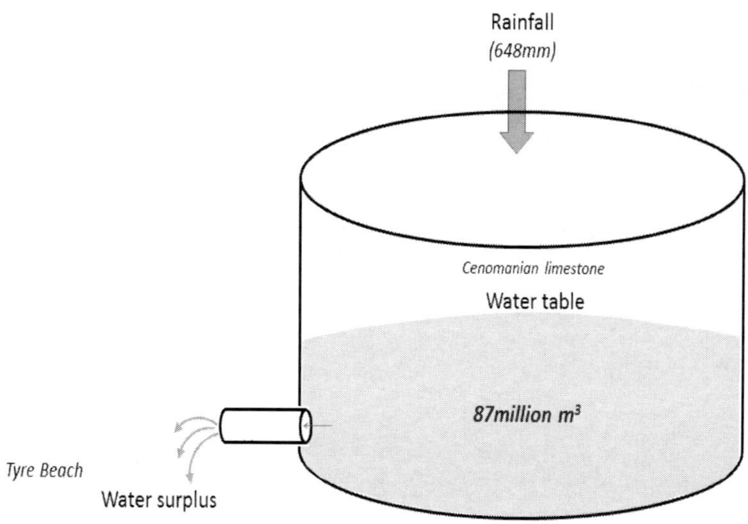

Figure 6.5. Scheme showing groundwater flow mechanism in Tyre Beach site.

Therefore, the declined rainfall rate (38 mm over the last three decades) accompanied with increased temperature at 1.4°C did not remarkably influence on the volume of water in the wetland site. This was evidenced by the analyzed stream flow in the three investigated flow meters fixed in the area of concern.

The majority of wetland components are well existed in this site, including diverse flora and fauna species (e.g., decreased in the number of one species of the migratory birds by about 20%), saturated soil with characteristics oriented towards saturated soil as proved from the textures and chemical components. In addition, water sources with different aspects occur. Therefore, this has been well characterized by the evaluation of wetland criteria which was estimated at about 74/100, as mentioned in the previous section.

However, negative human impact is also recorded in this site, but with least damage, and in the surrounding regions. This includes:

1. Frequent forest fires
2. Presence of large-scale landfill in the proximity of the site (Figure 6.6)
3. uncontrolled groundwater pumping
4. Excessive spread of fertilizers
5. Fragmentation of arable and cultivated lands, mainly by roads.
6. Uncontrolled urban expansion
7. Presence of contaminated effluents in the proximity of the site
8. Unmanaged irrigation systems (e.g., furrow irrigation, etc.).
9. Little diverting and management of water resources (e.g., lack of well-constructed channels, water loss to the sea).

Proposed Recommendations

Even thought, Ras Ech-Chekkaa site is much more under physical and anthropogenic stresses, yet both sites must be considered as natural reserves. Meanwhile, the sustainability of water resources in Tyre Beach site is attributed to two reasons:

1. The plentiful of water resources as mentioned above
2. Presence of Tyre Beach natural Reserve which results or controls on the wetland site itself.

Figure 6.6. Landfill in the proximity (<250 m) of Tyre Beach site.

The proposed recommendations are viewed from three levels of site conservation. These are: individual, institutional and legal levels as follows:

Individual Level

1. Awareness must be raised through different dissemination sources (e.g., workshops, awareness campaigns, media, etc.)
2. Ethics towards natural resources reservation, with a special concern to water resources and the unique flora and fauna
3. Rational water management on the individual level must be adopted, such as wise use of water, mini-projects for water harvesting and following efficient irrigation system

Institutional Level

1. Protection aspects must be considered for the designated sites. This includes (but not limited): building fences, water channels, water ponds
2. Water supply management approaches to reduce the stresses on the existing groundwater sources
3. Adopting site guards for the wetlands and the surrounding regions
4. Controlling different aspects of human activities in/around the proximity of the wetland sites

5. Applying detailed studies on wetlands of Lebanon, with a special emphasis to groundwater regime assessment and climate change impact
6. Including wetland areas in water strategies at the national level.
7. Re-empower efforts to include Lebanon's wetlands in the global and regional projects and conventions.

Legal Level

1. Establishing specified environmental and water legislations for Lebanon's wetlands, including the surrounding areas.
2. Activating proposed laws for hunting
3. Presuming water tariffs to mitigate the uncontrolled (and misused) water resources.

REFERENCES

AECID/MOE/IUCN. 2011. Supporting the Management of Important Marine Habitats and Species in Lebanon. *Survey of the Proposed Marine Protected Area at Chaka, Lebanon. Project draft.*

AECID/MOE/IUCN. 2011. Supporting the Management of Important Marine Habitats and Species in Lebanon. *Survey of the Proposed Marine Protected Area at Chekkaa, Lebanon. Project draft.*

CAL. *Atlas Climatique du Liban, Tome 1* (Climate Atlas of Lebanon, Volume 1). Service Météorologique, Ministère des Travaux publics et Transports (Meteorological Service, Department of Public Works and Transport). 1971, 45p.

CAL. *Atlas Climatique du Liban, Tome II* (Climate Atlas of Lebanon, Volume 2). Service Météorologique, Ministère des publics et Transports (Meteorological Service, Department of Public Works and Transport). 1980, 40p.

CAL. *Atlas Climatique du Liban 1973. Tome 1* (Climate Atlas of Lebanon 1973. Volume 1). Service Météorologique, Ministère des Travaux publics et Transports (Meteorological Service, Department of Public Works and Transport), 31p.

Chaigneau, A., Gizolme, A., Grados, C., 2008. Mesoscale eddies off Peru in altimeter records: identification algorithms and eddy spatio-temporal patterns (Mesoscale eddies off Peru in altimeter records: identification algorithms and eddy spatio-temporal patterns). *Prog. Oceanogr*. 79, 106-119.

CNRS. 1999. TIR survey for freshwater sources in the marine environment. *National Center for Remote Sensing. Final Report*. LNCSR. NCRS, 103p.

Cowardin, L., Carter, V., Golet, F., and La Roe, E. 1992. Classification of wetlands and deep water habitats of the United States. US Department of the Interior. WS/OBS-79/31. Fish and Wildlife Service, Office of Biological Services. Washington DC. 181p.

DAG. 1962. Direction des Affaires Géographiques et Géodésiques (Directorate of Geographic Affairs and Geodetic), Institution et rattachement au Ministère de la Défense Nationale-Commandement de l'armée libanaise (Institution and attachment to the Ministry of National Command of the Lebanese Army Defense).

Darwish T., Khawlie M., Jomaa I., Abou Daher M., Awad M., Masri T., Shaban A., Faour G., Bou Kheir R., Abdallah C. and Haddad T. 2006. *Soil Map of Lebanon 1/50000*. CNRS-Lebanon, Monograph Series 4, pp. 367.

DGAC. *Direction Générale de l'Aviation Civile, rapport annuel* (Directorate General of Civil Aviation, Annual Report), Beyrouth, Liban. 1999, 32p.

Dubertret, L. 1953. *Carte géologique de la Syrie et du Liban au 1/50000 me* (Geological Map of Syria and Lebanon to 1/50000 me). 21 feuilles avec notices explicatrices. Ministère des Travaux Publics (21 sheets with explicatrices notices. Ministry of Public Works). L'imprimerie Catholique, Beyrouth, 66p.

EcoMed (Ecology and Mediation International). 2012. *Green Belt Project Preliminary Biodiversity Assessment report La Cimenterie Nationale SAL Chekka, Lebanon*. By Cluchier, A., Fleury, S., Khater, C.

El-Khouri, T. 2012. Conservation Needs for Mediterranean Wetlands (with a focus on the Aammiq Wetland). Stable Institutional Structure for Protected Areas Management.

Hassan, Z. 2015. Impacts of climate change on Lebanese wetlands and their migratory soaring birds/legislative framework. Master dissertation. Lebanese Uinversity. 78p. http://docs.lib.noaa.gov/rescue/data_rescue_lebanon.html#o48764368.

LARI (Lebanese Agricultural Research Center). *Climatic Data. Monthly Bulletin*. 2003.

LRA. 1970. Litani river authority. Unpublished report. 1970; 23p.

LRA. 2012. Litani river authority. Unpublished report. 2012; 29p.

Ministry of Environment. 2006. MedWetCoast, United Nations Devleopment Program, and Fonds Français pour l'Environnement Mondial (French Fund for Global Environment). *Aammiq Wetland Management Plan.*

MOE/UNDP/LU (Lebanese University). 2004. Biodiversity Assessment And Monitoring In The Protected Areas/Lebanon Leb/95/G31 Tyre Beach Nature Reserve. *Final Report.* 144 pp.

NOAA. 2012. National Oceanographic Data Center. NOAA Central Library, RAC/SPA - UNEP/MAP. 2014. Ecological characterization of sites of interest for conservation in Lebanon: Enfeh Peninsula, Ras Chekaa Cliffs, Raoucheh, Saida, Tyre and Makoura. By Ramos-Espla A.A., Bitar G., Khalaf G., El Shaer H., Forcada A., Limam A., Ocaña O., Sghaier Y.R. and Valle C. Ed. RAC/SPA - MedMPAnet Project, Tunis. 168 pages + annexes.

Sanlaville, P. 1977. Etude géomorphologique de la région littorale du Liban (Geomorphological study of Lebanon's coastal region). *Publications de l'Université Libanaise. Section des études géographiques, Beyrouth, tome I* (Publications of the Lebanese University. Section of Geographical Studies, Beirut, Volume I), 405 p.

Shaban. 1985. Geology and Hydrogeology of the Nabatieh area. Master dissertation. American Uinversity of Beirut. 78p.

Shaban. A. 2011. Analyzing climatic and hydrologic trends in Lebanon. *Journal of Environmental Science and Engineering*, No. 3, Vol. 5, 2011.

Shaban. A., 2013. Geomorphological and geological aspects of wetlands in Lebanon. *3rd Int. Geography Symposium* (Geo-Med 2013). 10-13 June, 2013, Antalya - Turkey.

Shaban, A., 2014. Physical and Anthropogenic Challenges of Water Resources in Lebanon. *Journal of Scientific Research and Reports*. Vol. 3 Issue 3. 2014. 164-179.

Shaban, A., Houhou, R. 2015. Drought or humidity oscillations? The case of coastal zone of Lebanon. *Journal of Hydrology*. 529(2015):1768-75.

UNDP. 1970, Liban: *Etude des eaux souterraines* (Study of groundwater). Programme des Nations Unies pour le développement (United Nations Development Programme), N.Y. DP/SF/UN, 44, 185p.

UNEP (United Nations Environment Program). 2011. Proposal for inclusion in the SPAMI List: Tyre Coast Nature Reserve. *Mediterranean Action Plan*. Regional Activity Centre for Specially Protected Areas (RAC/SPA). 51 pp.

WRB (World Reference Base for Soil Resources). 2014. International soil classification system for naming soils and creating legends for soil maps. *World Soil Resources Reports 106*. FAO, 2014. 181 pp.

In: Coastal Zones
Editor: Gina Moran

ISBN: 978-1-63485-611-9

Chapter 3

EXPANSIONISM OF MARITIME BORDERS: BRIEF CONSIDERATIONS ON OCEAN GEOPOLITICS THROUGH THE ORIGINS OF ARTICLE 76 OF THE UNCLOS

Etiene Villela Marroni[1,*] and Ígor Villela Marroni[2,†]
[1]Laboratory of International Policy and Management of Oceanic Space,
Federal University of Pelotas, Pelotas, Brazil
[2]Laboratory of International Policy and Management of Oceanic Space,
Research Associate, Pelotas, Brazil

ABSTRACT

The diversity of uses of the oceanic space and the old conception of "freedom of the seas doctrine" has compelled an adaptation of the political-economic and spatial legal framework for the oceanic ecosystem. This changing of direction encompassed the international system and has given rise to a new geopolitics for the legal framework of oceanic spaces around the globe, in terms of the United Nations Convention on the Law of the Sea (UNCLOS). Every coastal state who has signed the Convention acquired the right to claim its underwater territory or extended continental shelf up to 350 nautical miles from its

* E-mail: etiene.marroni@ufpel.edu.br; etivm@terra.com.br.
† E-mail: imarroni@terra.com.br.

coast. This study aims to account for procedures taken by coastal, insular, and archipelagic states when claiming the extension of its oceanic limits.

Keywords: coastal zone, marine spatial planning, article 76 of the UNCLOS, ocean management

INTRODUCTION

It is imperative to consider the historical, political and socio-economic background in order to realise the importance of the study of the coastal States' continental shelf in the context of international politics. Recapitulating its process, it can be assumed that discussions began after the Second World War with US President Harry Truman's proclamation of 1945 and the creation of the United Nations.

In the period between 1949 and 1956 a new codification of the use of oceans was submitted to the International Law Commission. The Commission was set by the General Assembly of the United Nations in 1948 with the aim of preparing draft Conventions on matters that were not yet ruled by international law, as it was the case ofpç the political management of oceans. Thus the First United Nations Conference on the Law of the Sea was installed, setting out four parallel "Conventions" (Territorial Sea and the Contiguous Zone; High-Seas; Continental Shelf; and, finally, Fisheries and Conservation of High Seas Biological Ressources). The 1958 Convention on the Continental Shelf was an attempt to define criteria for the exploitation of the continental shelf by coastal States. At the time it was not clear whether the definition of legal continental shelf would include the whole continental margin. There were two definitions on the agenda: continental margin and continental shelf. In the Convention's terms, according to Cavnar (2009), the concepts were as following:

- Continental Margin is a geological concept of geographical and/or geomorphological nature that can be defined as a submerged zone of the continent gently sloping down from the baseline/coast;
- Continental Shelf is a geological concept which may possess a legal connotation used for demarcating the oceanic borders (limits) of a given coastal State.

Thus, it was resolved that the issue of the continental shelf would be debated in the next meeting in the city of Geneva (Switzerland). So the beginning of the studies for the Second United Nations Conference on the Law of the Sea of 1960 was settled. As Castro (1969) have pointed out, because of the Cold War the United States and the Soviet Union were absent from the Conference which was therefore nullified. In that opportunity no agreements were met on matters such as the width of the territorial sea, legal frameworks of fishery, neither on the study of the continental shelf. It must be added that Brazil did not ratify the first two Conventions.

Discussions went on between 1968 and 1970 on the platform of the Preparatory Commission for the Third United Nations Conference on the Law of the Sea. Since no consensus or regulations on oceanic management were agreed on, in 1970 a Statement of Principles was signed, in which countries commited to participate, in the following years, in discussions and forums for the regulation of the oceanic space. According to Cavnar (2009), what had began as a concern about deep sea ressources soon became a global diplomatic effort to regulate and create formalised rules in all areas concerning the oceans. These initiatives led into the Third United Nations Conference on the Law of the Sea aiming at a global treaty concerning the oceans. The Conference took place in Caracas, in 1973.

THE BEGINNINGS OF A GLOBAL OCEAN MANAGEMENT

According to Elferink (2005), in the period between 1958 and 1982 several multilateral meetings seeking a conceptual and definitive approach to the study of the continental shelf took place. The 1982 Convention, as provided by its Article 76, strove to establish substantive rules in order to ensure the limits of the coastal States' continental shelf defined in this study as new boundaries, which would not be subject of future changes.

There are countless issues pertaining Article 76 that certainly cannot be addressed in their totality. As we proceed to a general overview, we will consider two relevant issues: the first one concerns the political nature of Article 76, arising from a legal provision which should be interpreted and assimilated as a national and international public policy. The second issue regards the legal-scientific interface, brought about by the terms "legal continental shelf" and "scientific continental shelf."

Souza (2000) state that the Convention defines a detailed regulation framework of oceanic space, of the limits of national jurisdiction, of sea access, navigation, maritime environment protection, biological resources extraction and conservation, maritime scientific research, extraction of seabed mineral and other non-biological resources. Besides, it commits to dispute settlement. It establishes rights and obligations over oceanic spaces and regulates all related activities. This way, according to the authors, the coastal State is entitled to a Territorial Sea, a Contiguous Zone, an Exclusive Economic Zone and a Continental Shelf (in case this exists), all of which will be ruled by specific rights and jurisdictions. It is guaranteed, as well, that all states have rights and obligations regarding the extraction of mineral resources from the sea floor lying beyond national jurisdiction. Thus, a number of concepts were recognized, which are of legal nature and pertain domains beyond coastlines of continents, islands or archipelagos. These concepts include: Territorial Sea, Contiguous Zone, Exclusive Economic Zone and Continental Shelf and the International Seabed Area. Between 1973 and 1982 it was acknowledged that issues regarding the continental shelf (in particular concerning mineral resources) would lead to socio-economic implications while this concept simultaneously referred to a physical entity rather correlated to the geophysicist's scientific domain then to that of political scientists, economists or legal scholars. Such characteristics lead us to think that the process of establishing the continental shelf's limits (oceanic boundaries), nevertheless being a highly technical one, should as well be understood, according to Article 76, as a political-diplomatic process, conveyed in terms difficult to understand.

The Extended Continental Shelf should be considered building upon Article 76 that is based on the fact that the expansion of oceanic boundaries implies a series of negotiations and decision-makings, not only between states and international organizations, but also between neighbouring states whose sea beds may hold large mineral resources, which in turn inspire other nations' greed. In addition there is the question of the exact outer limits of the Continental Shelf, which sets up the limits of the International Area, a sensitive zone for foreign incursion by other coastal States.

The understanding of Article 76 from a political standpoint was made difficult by its diction, for it has been written in a highly specialised legal and scientific-technical language. Firstly, as stated by Persand (2005), it embodies scientific-technical elements (of geological, geophysical and geomorphological nature) which are essential for its correct application. Secondly, the Commission on the Limits of the Continental Shelf (CLCS), albeit deep rooted in politics, was set up as a technical body playing a key role in the process of tracing the outer limits of the Continental Shelf beyond 200 nautical miles (Extended Continental Shelf). The Commission (bound to the UN) is composed exclusively by scientific-technical experts (from the fields of Geology, Geophysics or Hydrography), a fact that sometimes hinders the understanding of its directives.

Key issues are still far from being answered, namely: in what way will the political nature of coastal States, when applying Article 76, ensure a transparent scientific-technical execution? In which degree are the scientific-technical terms of article 76 to be interpreted when the matter is to deal with a legal provision? With such "specific" terms, since the Commission on the Limits of the Continental Shelf's members are scientific-technical experts bound to the United Nations, how will a coastal State "adjust itself" in order to obtain an extension of its oceanic borders since the wording of Article 76 represents a compromise between several proposals presented during the Second United Nations Conference on the Law of the Sea? Likewise, insofar the CLCS is entitled to evaluation in the case of some scientific-technical cases, how to deal with such issues in the political and legal realms?

So, as we try to interpret Article 76, we distinguish between three aspects to be considered: the political aspect (involving bilateral and multilateral negotiations between countries with similar claims); the scientific-technical aspect (which represents the practical aspect of the expansion of oceanic boundaries); the legal aspect (which represents the formal, normative aspect of the political and practical processes). Therefore, the enlargement process of the Extended Continental Shelf or, in other words, the expansionistic process of coastal and island States into the sea depends on several interdisciplinary factors conditioned by the "hybrid" structure of Article 76.

We understand that the establishment of the outer limits of the Continental Shelf beyond 200 nautical miles in accordance with Article 76 of the Convention is a complex process that involves multidisciplinary aspects and demands a whole lot in terms of capacity, expertise and financial resources. In order to realise the inclusion of Article 76 into the Convention and its importance in oceanic geopolitics, it is indispensable to analyse the origins of Article 76.

UNDERSTANDING ARTICLE 76: BRIEF CONSIDERATIONS ON OCEAN GEOPOLITICS

Article 76 of the Convention consists of 10 paragraphs dealing with a variety of interconnected issues. In the first paragraph we find a definition of the legal Continental Shelf with two alternatives concerning the outer limits: 200 nautical miles from the baselines or up to the outer border of the continental margin, which extends beyond that distance. The Continental Margin, in turn, is defined in paragraph 3 as comprising the geophysical platform, the Continental Talus and the continental elevation. Paragraph 2, according to the general definition of continental shelf goes further, pointing out that the continental shelf should not go beyond the outer borders, which are specified in paragraphs 4 and 6 of the same Article. This clause is relevant for coastal States that have not yet determined the outer limits of the continental shelf in accordance with paragraph 8 of Article 76. We quote some relevant passages from Article 76:

> 2. The continental shelf of a coastal State shall not extend beyond the limits provided for in paragraphs 4 to 6.
>
> [...]
>
> 4. (a) For the purposes of this Convention, the coastal State shall establish the outer edge of the continental margin wherever the margin extends beyond 200 nautical miles from the baselines from which the breadth of the territorial sea is measured, by either:
>
> (i) a line delineated in accordance with paragraph 7 by reference to the outermost fixed points at each of which the thickness of sedimentary rocks is at least 1 per cent of the shortest distance from such point to the foot of the continental slope; or
>
> (ii) a line delineated in accordance with paragraph 7 by reference to fixed points not more than 60 nautical miles from the foot of the continental slope.
>
> (b) In the absence of evidence to the contrary, the foot of the continental slope shall be determined as the point of maximum change in the gradient at its base
>
> [...]
>
> 6. Notwithstanding the provisions of paragraph 5, on submarine ridges, the outer limit of the continental shelf shall not exceed 350 nautical miles from the baselines from which the breadth of the territorial sea is measured. This paragraph does not apply to submarine elevations that are natural components of the continental margin, such as its plateaux, rises, caps, banks and spurs (UNCLOS, 1982).

In order to better grasp the complexity involved in Article 76 it is necessary to mention that at the beginning of the negotiations for the Third United Nations Conference on the Law of the Sea, one of the tasks posed to participants was the precise definition of the seabed area and its subsoil beyond the national jurisdiction of coastal States. By the end of the 1960s this issue became a matter of urgency as it became clear that the extraction of seabed mineral resources on a commercial scale would become possible in a near future. From a technological standpoint, given the international scenario in those times, those activities would mainly benefit developed nations. However, developing countries were successful with their proposal that resources from the seabed and its subsoil beyond the limits of national jurisdiction would be considered common heritage of humanity. If it had been the case that the outer maritime boundaries of all the coastal States had been established from a certain distance from the coast, by the time of the Conference, there would be no reason for further studies in search for specific criteria, because there existed a consensus on the necessity of a clear and definitive distinction between areas of national jurisdiction and international areas.

It is unlikely that, prior to the negotiatons that lead to Article 76, the legal continental shelf would have been understood as the length up to the outer border of the continental margin, as it was presented by nations detaining a large margin (or extended continental shelf). Nevertheless, some nations were expecting already that the legal continental shelf would stretch out beyond 200 nautical miles in some parts of the world. The 200 miles mark was advocated as the limit between coastal States national jurisdiction areas and the ocean international area. Most negotiators worried about conciling positions that would result in a definitive solution of border demarcation. Article 76 was able to combine the points of view of States that possessed an extended continental shelf. An important concession was made in Article 82 of the Convention, which provides that coastal States, through the International Seabed Authority, shall pay contributions to the international community regarding the utilization of non-living resources of the continental shelf beyond 200 nautical miles.

Since less developed States aimed at establishing the international seabed beyond national jurisdiction as "common heritage of humanity," the "exploitation criterion" of the 1958 Convention had to be revised during the negotiations at the Third United Nations Conference on the Law of the Sea. In fact, one of the basic goals was a precise definition of the extent of maritime borders of coastal States. Back then, the text of Article 76 was elaborated in order to guarantee coastal States the right to the extended continental shelf without altering the concept of common heritage of humanity.

One aspect of Article 76 that causes controversy is its highly technical-scientific terminology dealing with the limits of the extended continental shelf.

Its guidelines incorporate a series of terms "imported" from Geosciences, a fact that has triggered discussions concerning its correct interpretation. So the Convention brought the Commission on the Limits of the Continental Shelf (CLCS/UN) into light, a specific body aiming to support Coastal states in establishing the said limits.

Considering the Convention as an international public policy, it is supposed that all signatory coastal States have to regulate their own maritime national public policies in accordance to the considerations expressed in the final document, bringing law into line with its requirements as well. Thus, according to Antunes and Pimentel (2003), those States would be investing in professional development of experts which then would be able to apply the contents of the Convention in their countries of origin. The political interpretation of the Convention demands the work of a multidisciplinary team. It is assumed that not all Treaty parties are prepared to deal with this new reality, particularly with the implementation of Article 76, what causes "external" dependency of more disadvantaged coastal States (in this case, the consultation of professionals from the CLCS) when solving problems of national interest.

Persand (2005), aware of the political problems generated by the complexity involved in the implementation of Article 76, observed that right before the start of negotiations two limitations were considered for the expansion of continental shelves of coastal nations. The first one resulting from the fact that some nations possessed a Continental Shelf-Wide with a pronounced descending slope lightly inclined from the continental baseline. The sediments that form this continental increase come from the adjacent continent. This type is usually found on the Atlantic coast while the other type does not display such a simple pattern. Coastal nations with seismic activity and located in zones of fragmentation of the continental crust come into the picture in this context. They display a set of sedimentary basins or a narrow continental shelf. Their borders to the sea are usually formed by steep slopes leading to the deep seabed. This second kind is often found in the Pacific Zone.

Carleton et al. (2000) have pointed out, for this same context, that on the course of negotiations, the limits of the extended continental shelf were to be established based on the water depth, on natural boundaries of the submerged portion of the territory and by the borders of the continental crust. However, they have alerted to the fact that no agreement could be made based on such parameters precisely because the morphology of coastal States may vary substantially. They explain that most studies on this topic published by the time of the deliberations were under the responsibility of the Woods Hole Oceanographic Institution and the Lamont-Doherty Earth Observatory, both located in the United States of America. They added that those involved in the writing of Article 76 provided data about the extended continental shelf containing morphological profiles which were exaggeratedlly vertical, displaying a steep inclination downwards. The authors pointed out that this might have given the impression that all continental margins of coastal nations were equal, and therefore, the mapping of any of its aspects would have been a relatively simple task.

We come to the conclusion that it is important first to learn about the diversity of spaces of coastal, island and archipelagic States beyond the baselines legally acknowledged by the Convention and subsequently solve the issue of demarcating the extended continental shelf.

The legal demarcation of maritime spaces by the 1982 Convention was necessary for the organization of the use of the oceanic ecosystem. States are entitled to the rights and duties from the proper management of such territories. According to Van Pay (2008), the interest in delimitating maritime spaces by coastal States possessing an extended continental shelf and the resulting disputes arise from their sovereign rights over the exploitation of ocean resources. Coastal States are burdened with high expenses in order to prove the legality of the expansion of their borders (increase of the continental shelf). Currently more then 60 coastal States are interested in this new oceanic space and willing to proclaim and demarcate their sovereign territory. The concrete definition of these rights also in geographical and political terms is of paramount importance for the security, protection and utilization of resources derived from the extended continental shelf.

It is clear that the competition for the extended continental shelf is, above all, an issue of sovereignity related to the expansion of physical domains containing large natural resources. We call attention to the socio-economic and environmental-political implications of such expansions for nations, for the outer limits of the extended continental shelf will constitute, at the same time, the limit to the international area. The new borders will be a sensitive area to foreign invasion.

DEFINITION OF THE CONTINENTAL SHELF AND CRITERIA TO ESTABLISH ITS OUTER LIMITS

According to Article 76 paragraph 1, with the title "Definition of the continental shelf":

> The continental shelf of a coastal State comprises the seabed and subsoil of the submarine areas that extend beyond its territorial sea throughout the natural prolongation of its land territory to the outer edge of the continental margin, or to a distance of 200 nautical miles from the baselines from which the breadth of the territorial sea is measured where the outer edge of the continental margin does not extend up to that distance (UNCLOS, 1982).

This is a legal definition and should not be confused with the scientific definition.

The most difficult task is to specify the process that would establish a precise definition of the continental shelf's outer limits, precisely because of the interests at stake. These would hamper the application of a single generally aplicable standard. The resulting compromise arises from a serious approach aiming to meet all the interests involved.

Smith and Taft (2000) believe that the term "continental shelf" has a scrictly legal meaning. They refer to a United Nations study on the continental shelf that confirms this legal concept with strong political connotations in its relationship to the physical aspect of a natural prolongation of the continent, resulting in extension of the borders of coastal States.

Cavnar (2009) argues that the determination of the outer limits beyond 200 nautical miles is a particularly sensitive process that demands a considerable research period. The absence of defined outer limits has consequences in terms of sovereign rights over continental shelf areas beyond that limit and could trigger border conflicts between coastal nations.

Antunes and Pimentel (2003) assert that studies of the continental shelf have the purpose of assuring sovereign rights over the use of natural resources in waters, soil and subsoil, particularly over mineral resources and fossil fuels like oil and natural gas. Natural resources of the continental shelf comprise mineral and other non-living resources from the seabed and subsoil as well as living organisms. According to the Convention, a coastal State exercises sovereign rights over the continental shelf for the purpose of exploitation and utilization of natural resources.

A conflict of interests over legal or sovereign spaces actually took place in the course of the preliminary meetings leading to the Convention. Several issues related mainly to economic factors were raised. The less jurisdiction or sovereignity a coastal State had over the oceans, the greater would be the possibility of appropriation of the related areas through other States possessing the relevant technology. Martins (2009) sees sovereignity as one the key characteristics defining a State governed by the rule of law. One aspect of sovereignity is the State's jurisdiction over all those who are submitted to it. Therefore, a State shall not submit itself to any jurisdiction other than its own. The maritime territory, i.e., the waters under the jurisdiction of a certain State emcompass the territorial sea, the contiguous zone, the exclusive economic zone and the continental shelf. Hence, there is no reason to mix up maritime territory with territorial sea.

Antunes and Pimentel (2003) make a stand on the fact that the sovereign right of a coastal state over its extended continental shelf is based upon the lengthening of its submerged portion to the high seas. This right is represented by the natural prolongation of the continental margin. The fact that Article 76 contains a brief definition of continental shelf as well as specific rules for the definition of its outer limits confirms the right to determine the outer limits, provided that lines of restriction are observed.

The setting of the outer limits of the continental shelf, in accordance to Article 76, provides a coastal State the rights of exerting sovereignity over parts of its continental margin that do not surpass what are the potential lines of the outer limit. However, it must remain clear that the continental shelf shall not extend beyond the limits set by the Convention.

CONCLUSION

Taking the aforementioned considerations into account, the importance of Article 76 for the delimitation of new oceanic areas by coastal States becomes clear. The increase of a State's territory, in this case, demands a standardised procedure of multidisciplinar nature and that is difficult to implement, particularly for disadvantaged countries.

According to the directives set by the Commission on the Limits of the Continental Shelf (CLCS/UN), only the coastal State itself is empowered to implement such procedure. The absence of outer limit lines beyond the 200 nautical miles raises doubts on the exact range of the continental shelf, making it difficult for the coastal State to exercise rights of sovereignity over the areas.

In short, because of the ambiguity and vagueness of Article 76, the assessment of its political and legal content requires a holistic interpretation. Hence, it shall take the final text into account, as well as the object and aim of the Convention, underlining all aspects of its preliminary meetings, the mutual relations between its paragraphs and methodological rules of interpretation.

In this perspective, it can be said that the prototypical morphology of the continental margin rarely occurs given the variety of geomorphologic shapes resulting from different tectonic configurations. This approach is adopted also by the directives of CLCS/UN, which state: "subdivision implemented for the calculation of shelf margins, inclination and elevation may not always exist due to the variety of geological and morphological types of continental margins resulting from different tectonic and geological scenarios."

Antunes and Pimentel (2003) warn that Article 76 contains terms permeated by "scientific-political dualism." The CLCS/UN recognizes that some scientific-technical terms incorporated by the Convention originate in geo-morphological terms used in a complex diplomatic-legal-political context that sometimes edges away from scientific terminology and definitions that need to be understood by all parties involved in the Convention. It is important to understand such terminologies in order to take the process of broadening of maritime borders forward. Besides, the implementation of Article 76 may result in two kinds of rulings: first, the "legal continental margin" shall not reach the border of the continental elevation (i.e., a portion of the continental margin will stay outside of national jurisdiction); second, the "legal continental margin" may surpass the border of the continental elevation (i.e., a portion of the deep seabed becomes part of the legal continental shelf of a coastal State). In order to clarifiy this ambiguity in the same legal ruling it is necessary to clearly distinguish between geo-scientific and juristic-political concepts.

When looking at the issue from this standpoint it becomes clear that principles for political interpretation are connected to the scientific and legal aspects of the United Nations Convention on the Law of the Sea (UNCLOS). It can therefore be assumed that all countries interested in the increase of their maritime jurisdictions may submit themselves to the normative rulings in a consistent way. In short, the Convention should be interpreted and applied in a way that leads to practical effects.

Article 76, in spite of its terminological ambiguity in the "scientific-political" sense, carries two proposals for the determination of the extended continental shelf. Therefore, terminology originated from geosciences should be analysed with caution for it may have two distinct meanings: a geo-scientific and a legal-political meaning. The debate over it surely has to take this into account.

Its interpretation is an interdisciplinary exercise, for it is neither scrictly political, nor legal or scientific-technological, but it may progress inside a political-legal-scientific interface. Probably in most cases both types of issues – political and scientific-technological aspects – might be treated separately. Yet there are problems where the political, legal and scientific-technological aspects are so interconnected that analysing them separately is rendered almost impossible. A narrow study, be it from a political, legal or a scientific-technical standpoint will not be able to emphasize the importance of Article 76 in the context of expansionism resulting from the advancing of national sovereignities over seas and oceans. This aspect can be considered as being a hybrid policy envolving issues from the political as well as from the scientific field.

The determination of the morphological features of an oceanic crust, ridge or a sea mount is a scientific problem. Once these geographical factors are resolved, the juridical component comes into play: legalizing the area as "definitive and biding" by the submitting coastal State. At this point, the political negotiation process takes place in the windings of scientificism in order to secure a certain territory to a country.

Yet political, legal and scientific considerations surely should not be treated as equal. It is necessary to recognise that technical and political concepts and interpretations shall not pass beyond legal limits enforced by Article 76. This is because Article 76 is part of the UNCLOS which in turn is the result of a great political-diplomatic effort involving several countries. This effort was done in order to meet a legal provision that would take into account interests and necessities of all countries of the globe (even those without coastlines) and that therefore, it should be observed and respected.

In summary, the scientific-technological aspects are requirements for the demarcation of oceanic limits lawfully guaranteed. Thus, the exercise of technological-scientific discretionary judgement should not pass beyond the limits set by the CLCS/UN, a commission created under the auspices of the UNCLOS and legally acepted by all the Convention's signatory parties. The CLCS/UN, although being a body with very specific features is also a genuine decision-making body. Its members cannot set directives in disagreement with what was established in the summary of the negotiations of the Convention nor can they invoke rules that restrict the sovereign freedom of States without the consent of the Convention. It must be clear that the framework of interpretation and implementation of such a comprehensive Treaty as the UNCLOS is an exercise in diplomatic-political relationships between States.

The real benefit guaranteed by Article 76 is the possibility it gives coastal States to define new oceanic borders. However difficult the interpretation of its formulations may be, there are no doubts that a limit may be determined or, better still, it can be regulated in conformity with Article 76.

References

Antunes, N. M.; Pimentel, F. M. (2003). Reflecting on the Legal-Technical Interface of Article 76 of the LOSC: Tentative Thoughts on Practical Implementation. *ABLOS Conference Addressing Difficult Issues in UNCLOS*. <http://www.gmat.unsw.edu.au/ablos/ABLOS03Folder/PAPER3-1.PDF>.

Carleton, C. M.; Shipman, S.; Monahan, D e Parson, L. (2000). The practical realization of the continental shelf limit. In: Cook, P.; CARLETON, C. (Ed.). *Continental Shelf Limits*. The Scientific and Legal Interface. Oxford: Oxford University Press.

Castro, R. N. L. de. (1969). Aspectos Fundamentais da Doutrina Brasileira sobre Plataforma Continental. *Revista Brasileira de Política Internacional*. Rio de Janeiro, ano XII, n. 47/48, set./dez.

Cavnar, A. (2009). *Accountability and the Commission on the Limits of the Continental Shelf*: deciding who owns the ocean floor. Institute for International Law and Justice. New York: University School of Law (mimeo.).

Elferink, A. G. (2005). Article 76 of the LOS Convention on the Definition of the Continental Shelf: Questions concerning its Interpretation from a Legal Perspective. *Netherlands Institute for the Law of the Sea*, Faculty of Law, Utrecht University. <http://www.iflos.org/media/9415/abstracts.pdf>.

Marroni, E. V. (2013). *Política Internacional dos Oceanos*: caso brasileiro sobre o processo diplomático para a plataforma continental estendida. Tese de Doutorado. Programa de Pós-Graduação em Ciência Política. Universidade Federal do Rio Grande do Sul.

Martins, E. M. O. (2009). *Soberania e Jurisdição Marítima*. <http://jus.com.br/revista/texto/13984/soberania-e-jurisdicao-maritima>.

Persand, S. (2005). *Project Officer. Mauritius Oceanography Institute*. United Nations. The Nippon Foundation of Japan Fellow (mimeo.).

Smith, R. W.; Taft, G. (2000). Legal Aspects of the Continental Shelf. In: Cook, P.; Carleton, C. *Continental Shelf Limits. The Scientific and Legal Interface*. Oxford: Oxford University Press. 2000.

UNCLOS (1982). United Nations Convention on Law of the Sea. <www.un.org/depts/los/convention_agreements/.../unclos/unclos_e.pdf>.

Souza, K. G. (2000). Recursos minerais marinhos além das jurisdições nacionais. *Rev. Bras. Geof.*, vol.18, n.3, p.455-465. <http://www.scielo.br/scielo.php?script=sci_arttext&pid=S0102-261X2000000300017&lng=pt&nrm=iso&tlng=pt>.

Van Pay, B. (2008). Why Does Defining the U.S. Extended Continental Shelf Matter? *Bureau of Oceans and International Environmental and Scientific Affairs.* <http://state.gov/index.php/entries/defining_extended_continental_shelf/>.

BIBLIOGRAPHY

Climate change and agriculture in Jamaica: agricultural sector support analysis / Ramasamy Selvaraju ... [et al.]. LCCN: 2014395437 Published/Created: Rome: Food and Agriculture Organization of the United Nations, 2013. Description: xvii, 108 p.: col. ill., maps; 25 cm. Links: http://www.fao.org/docrep/018/i3417e/i3417e.pdf ISBN: 9789251078945 9251078947 LC classification: S600.7.C54 C565 2013 Related names: Selvaraju, R. Food and Agriculture Organization of the United Nations. Abstract: Climate change is likely to have adverse effects on the agriculture sector in Jamaica. Increase in the intensity and frequency of climate-related natural hazards, escalating rainfall variability, droughts and floods combined with fragile ecosystems and coastal zones, and agriculture-dependent livelihoods all contribute to Jamaica's overall vulnerability to climate change. Jamaica's Vision 2030 highlights these vulnerabilities as a threat to sustainable development, and provides the framework within which corrective and preventive actions should be undertaken. Climate change adaptation is one of the outcomes of the Vision 2030 Jamaica-National Development Plan. In the context of this plan, there is need for more analysis of the likely climate change impacts in the short and medium term, along with identification of agricultural adaptation strategies. These need to be understood in order to inform policy interventions and agriculture support institutions. Within this context, the objectives of the study are: (i) to review current knowledge on vulnerability, past trends in climate, and impacts of climate variability and change on agriculture sector, and (ii) to explore technical and policy alternatives in order to cope with and adapt to impacts of climate variability and change more effectively. The study identified what the potential impacts are, considered what interventions are appropriate, and if and where they should occur. The scope of the study focused on broader policy directions and investment priorities in relation to climate change adaptation. The first two chapters of this book present overall background on the agriculture sector and vulnerability context. Chapter 2 specifically presents vulnerability of agro-ecosystems and food production systems in both temporal and special dimensions. Chapter 3 elaborates on the nature of climate variability and expected future changes in climate. The past trends in climate were described based on observation, analysis and also local perception of the communities. Chapter 4 presents the results of analysis of past extreme climate events on agriculture, ecosystems, biodiversity, and fisheries and aquaculture. This chapter also gives an overview of the impacts of climate change on fragile ecosystems, crop suitability, land degradation, water resources, biodiversity, fisheries and aquaculture, and the economy. Efforts to advance climate change adaptation should focus on support for decision-making and capacity building processes. These shape implementation of location-specific traditional and improved adaptation practices aimed at reducing the impacts of climate variability and climate change. In particular, this process of adaptation needs to address the needs of marginalized groups that are most vulnerable to climatic and socio-economic changes. Chapter 5 presents several coping strategies adopted by the local communities. It also proposes a series of types of adaptation strategies that are relevant to address multiple risks of climate change. The adaptation typologies range from assessment of vulnerability and risks to strengthening data collection, monitoring and forecasting, and promoting agricultural linkages to supply chains, weather risk insurance, and research and development linkages. Institutional capacity building and organizational networking with clear definitions of roles and responsibilities are essential to advance adaptation efforts. Chapter 6 elaborates the institutional and policy contexts to respond to climate change. This chapter highlights the importance of mainstreaming of climate change concerns into agriculture policies, programmes and plans.-- Source other than Library of Congress. Subjects: Climatic changes--Jamaica. Crops and climate--Jamaica. Notes: Includes bibliographical references (p. 81-85). Additional formats: Also available on the World Wide Web. Series: Environment and natural resources management series; no. 21

Climate change and global sustainability: a holistic approach / edited by Akimasa Sumi, Nobuo Mimura and Toshihiko Masui. LCCN: 2010030724 Published/Created: Shibuya-ku, Tokyo; New York: United Nations University Press, c2011. Description: xx, 316 p.: ill.; 24 cm. ISBN: 9789280811810 (pbk.) 9280811819 (pbk.) LC classification: QC903 .C546 2011 Related names: Sumi, Akimasa. Mimura, Nobuo, 1949- Masui, Toshihiko. Contents: Introduction: From climate change to global sustainability / Akimasa Sumi and Nobuo Mimura -- Structuring knowledge of climate change / Ai Hiramatsu -- Climate change risk communication / Seita Emori -- Overview of climate change impacts / Nobuo Mimura -- Climate change impacts in Japan / Yasuaki Hijioka -- Water resource problems due to climate change / So Kazama -- Climate change and world food production in the twenty-first century / Hiroyuki Kawashima -- Effects of climate change and global warming on marine ecosystems and fishery resources / Yasuhiro Yamanaka and Masahiko Fujii -- Vulnerability of coastal zones in the twenty-first century / Hisamichi Nobuoka and Satoshi Murakami -- Adaptation and mitigation strategies in response to climate change / Makoto Tamura and Nobuo Mimura -- New international framework beyond the Kyoto Protocol / Hiroshi Hamasaki and Tatsuyoshi Saijo -- Emission reductions policy mix: industrial sector greenhouse gas emission reductions / Seiji Ikkatai -- Shell energy scenarios to 2050 / Shell International BV -- Climate policy and international development cooperation / Shunji Matsuoka -- Induction to a low-carbon city - innovation of urban form and human activities / Keisuke Hanaki -- The process of political decision-making on climate change and journalism in Japan / Tokuhisa Yoshida -- The conceptions of "environment" and eco-philosophy / Hideo Kawamoto -- Integration of a low-carbon society with a resource-circulating and nature-harmonious society / Toshihiko Masui -- Future vision towards a sustainable society / Akimasa Sumi. Subjects: Climatic changes. Climatic changes--Social aspects. Global warming--Social aspects. Sustainable development. Notes: Includes bibliographical references and index. Series: Sustainability science series

Coastal zones: solutions for the 21st century / Juan Baztan, Omer Chouinard, Bethany Jorgensen. Edition: 1st edition. LCCN: 2015932108 Personal name: Baztan, Juan. Published/Produced: Waltham, MA: Elsevier, 2015. ISBN: 9780128027486

Coastal zones and climate change: risk and response / David Michel, Amit Pandya, editors. LCCN: 2011499575 Published/Created: Washington, DC: Henry L. Stimson Center, 2010. Description: xii, 106 p.: ill., maps; 26 cm. ISBN: 9780982193556 (pbk.) 0982193556 (pbk.) LC classification: QC982.5.I5 C63 2010 Related names: Michel, David 1966- Pandya, Amit A. Henry L. Stimson Center. Henry L. Stimson Center. Regional Voices. Regional Centre for Strategic Studies (Colombo, Sri Lanka) Subjects: Climatic changes--Effect of human beings on--Indian Ocean Region. Climatic changes--Indian Ocean Region. Climatic changes--Government policy--Indian Ocean Region. Coasts--Indian Ocean Region. Notes: Part of Stimson's Regional voices: transnational challenges project. "The work for the volume had its inception in a workshop cohosted by Stimson and the Regional Centre for Strategic Studies (RCSS) in Colombo, Sri Lanka, March 19-20, 2009"--P. x. Includes bibliographical references.

Development disparity between coastal and non-coastal zones / Dr. Rabindra Garada. Edition LCCN: 2014362403 Personal name: Garada, Rabindra, author. First edition. Published/Produced: New Delhi: SSDN Publishers & Distributors, 2014. Description: xxiii, 296 pages: map (black and white); 23 cm ISBN: 9789383575022 9383575026 LC classification: HC437.O7 G37 2014 Subjects: Economic development--Regional disparities--India--Odisha. Coastal zone management--India. Odisha (India)--Social conditions--Regional disparities. Odisha (India)--Economic conditions--Regional disparities. Notes: Includes bibliographical references (pages 279-294) and index.

EWATEC-COAST, technologies for environmental and water protection of coastal zones in Vietnam: contributions to 4th International Conference for Environment and Natural Resources--ICENR 2014, 17-18 June 2014, Ho-Chi-Minh City, Viet Nam / eds., G. Meon et al. LCCN: 2015398912 Meeting name: International Conference for Environment and Natural Resources (4th: 2014: Ho Chi Minh City, Vietnam), author. Edition: 1. Auflage. Published/Produced: Göttingen: Cuvillier-Verlag, 2014. Description: 220 pages: illustrations (chiefly color); 21 cm. ISBN: 9783954048526 (alk. paper) LC classification: TD313.V5 I58 2014 Related names: Meon, G. (Günter), editor. Subjects: EWATEC-COAST (Project)--Congresses. Water-supply--Vietnam, Southern--Management--Congresses. Environmental protection--Vietnam, Southern--Congresses.

Coastal ecology--Vietnam, Southern--Congresses. Notes: Includes bibliographical references. Series: HYWAG-Schriftenreihe, 2363-7218; Band 1

Geophysical hazards: minimizing risk, maximizing awareness / Tom Beer, editor. LCCN: 2009938629 Published/Created: Dordrecht; London: Springer, c2010. Description: xix, 262 p.: ill. (some col.), maps (some col.); 27 cm. Links: Table of contents only http://www.loc.gov/catdir/enhancements/fy1506/2009938629-t.html Publisher description http://www.loc.gov/catdir/enhancements/fy1506/2009938629-d.html ISBN: 9789048132355 (cased) 9048132355 (cased) 9048132363 (ebook) 9789048132362 (ebook) 9789400731400 940073140X LC classification: GB5014 .G485 2010 Related names: Beer, Tom. Summary: This volume brings together some of the most influential international scientists dealing with natural hazards and society's response to natural hazards. It recognises that hazards, like poverty, will always be with us but that concerted and organized societal response can prevent a hazard becoming a disaster. Contents: The hazards theme of the International Year of Planet Earth / Tom Beer -- Social science perspectives on hazards and vulnerability science / Susan L. Cutter -- Focusing on the environment and human security nexus / Juan Carlos Villagrán de León and Janos J. Bogardi -- Communicating geological hazards: educating, training and assisting geoscientists in communication skills / David Liverman -- Introduction of a new international research program: integrated research on disaster risk -- the challenge of natural and human-induced environmental hazards / Gordon A. McBean -- Building a university network for disaster risk reduction in sub-Saharan Africa / Genene Mulugeta -- Co-operation plan on hazards & disasters risk reduction in Asia and the Pacific / Harsh Gupta -- Closing the gap between science and practice to reduce human losses in hydro-meteorological disasters / Kuniyoshi Takeuchi -- The role of geosciences in the mitigation of natural disasters: five case studies / S.A.G. Leroy ... [et al.] -- Seismic hazard in India: practical aspects and initiatives during IYPE / R.K. Chadha -- Computational geodynamics as a component of comprehensive seismic hazards analysis / Alik Ismail-Zadeh -- Hazards in the coastal zones related to groundwater-seawater interaction processes / Y.A. Kontar, Yu.R. Ozorovich, And A.T. Salokhiddinnov -- Mega tsunami of the world oceans: chevron dune formation, micro-ejecta, and rapid climate change as the evidence of recent oceanic bolide impacts /

Viacheslav Gusiakov ... [et al.] -- Understanding slow deformation before dynamic failure / G. Ventura ... [et al.] -- Landslides in mountain regions: hazards, resources and information / Raisa Gracheva and Alexandra Golyeva. Subjects: Natural disasters--Risk assessment. Natural disasters--International cooperation--Research. Notes: Includes bibliographical references and index. Series: International year of planet Earth

Groundwater in the coastal zones of Asia-Pacific / Christoph Wetzelhuetter. LCCN: 2013939191 Personal name: Wetzelhuetter, Christoph. Published/Produced: New York: Springer, 2013. Links: Publisher description http://www.loc.gov/catdir/enhancements/fy1314/2013939191-d.html Table of contents only http://www.loc.gov/catdir/enhancements/fy1314/2013939191-t.html ISBN: 9789400756472

Handbook of ecological indicators for assessment of ecosystem health / edited by Sven E. Jørgensen, Fu-Liu Xu, Robert Costanza. LCCN: 2009051368 Edition: 2nd ed. Published/Created: Boca Raton, FL: CRC Press/Taylor & Francis, c2010. Description: xiii, 484 p.: ill., maps; 25 cm. ISBN: 9781439809365 (hard back: alk. paper) 1439809364 (hard back: alk. paper) LC classification: QH541.15.E265 H36 2010 Related names: Jørgensen, Sven Erik, 1934- Xu, Fu-Liu. Costanza, Robert. Summary: "Continuing in the tradition of its bestselling predecessor, the Handbook of Ecological Indicators for Assessment of Ecosystem Health, Second Edition brings together world-class editors and contributors who have been at the forefront of ecosystem health assessment research for decades, to provide a sound approach to environmental management and sustainable development. Significantly updated and expanded, this authoritative resource details a proven framework for selecting, evaluating, and validating ecological indicators for ecosystem health assessment. It guides readers through the application of this framework to a wide range of ecosystems, including wetlands, estuaries, coastal zones, lakes, forests, marine ecosystems, lagoons, agricultural systems, landscapes, and rivers. The text synthesizes material from a variety of books, journals, and private research, to consider biodiversity, energy needs, ecological economics, and natural capital in the measurement of ecological health. Organized for ease of reference, the first part of the handbook provides the required theoretical background. It presents a complete overview of all relevant ecological indicators - including thermodynamics, resilience estimates, exergy, and emergy indicators. The second part focuses on how to effectively apply the ecological indicators to a number of important ecosystems. It includes many examples and case studies that clearly illustrate the advantages and disadvantages of each method for specific applications. Offering first-hand insight and practical guidance from practitioners in the field, this complete resource supplies the tools and the well-rounded understanding required to diagnose the health of virtually any ecosystem with much improved accuracy."-- Publisher's description. Contents: Application of indicators for the assessment of ecosystem health / S. E. Jørgensen ... [et al.] -- Eco-exergy as ecological indicator / S. E. Jørgensen -- Emergy indices of biodiversity and ecosystem dynamics / Mark T. Brown and Sergio Ulgiati -- Eco-exergy to emergy flow ratio for the assessment of ecosystem health / F. M. Pulselli, C. Gaggi, and S. Bastianoni -- Natural capital security/vulnerability related to disturbance in a panarchy of social-ecological landscapes / Nicola Zaccarelli, Irene Petrosillo, and Giovanni Zurlini -- Species richness in space and time as an indicator of human activity and ecological change / Erich Tasser ... [et al.] -- Landscape development intensity and pollutant emergy/empower density indices as indicators of ecosystem health / Mark T. Brown and Kelly Chinners Reiss -- Ecosystem services and ecological indicators / Robert Costanza -- Application of ecological indicators for the assessment of wetland ecosystem health / S. E. Jørgensen -- Application of ecological indicators for the assessment of ecosystem health in estuaries and coastal zones / João C. Marques -- Application of ecological and thermodynamic indicators for the assessment of ecosystem health of lakes / Fu-Liu Xu -- Application of ecological indicators in forest management in Africa / Kouami Kokou, Adzo Dzifa Kokutse, and Kossi Adjonou -- Using ecological indicators to assess the health of marine ecosystems: the North Atlantic / Villy Christensen and Philippe Cury -- Indicators for the management of coastal lagoons: Sacca di Goro case study / J. M. Zaldívar ... [et al.] -- Ecosystem indicators for the integrated management of landscape health and integrity / Felix Müller and Benjamin Burkhard -- Integrated indicators for evaluating ecosystem health: an application to agricultural systems / V. Niccolucci ... [et al.] -- Ecological indicators to assess the health of river ecosystems / Carles Ibáñez ... [et al.]. Subjects: Ecosystem health. Environmental indicators. Notes: Includes bibliographical references and index. Series: Applied ecology and environmental management

Remote sensing of coastal environments / edited by Yeqiao Wang. LCCN: 2009035957 Published/Created: Boca Raton, FL: CRC Press/Taylor & Francis, c2010. Description: xviii, 423 p., [16] p. of plates: ill. (some col.), maps (some col.); 25 cm. ISBN: 9781420094411 (hardcover: alk. paper) 1420094416 (hardcover: alk. paper) LC classification: QH541.15.R4 R45 2010 Related names: Wang, Yeqiao. Summary: "As coastal environments around the world continue to face unprecedented natural and anthropogenic threats, recent developments in the technologies that support geospatial data acquisition, imaging, and computing have profoundly enhanced monitoring capabilities in coastal studies. Providing systematic treatment of the key developments, Remote Sensing of Coastal Environments brings together renowned scholars to supply a clear presentation of the state-of-the-art in this technically complex arena. Edited by a recipient of the prestigious PECASE award, this book provides unrivaled coverage of the issues unique to coastal environments. It presents the best available data for measuring and monitoring coastal zones and explains how decision makers and resource managers can use this data to address contemporary issues in coastal zone management. The text illustrates the latest developments in active remote sensing, hyperspectral remote sensing, high spatial resolution remote sensing, the integration of remote sensing and in situ data, and covers the effects of land-cover and land-use change on coastal environments. Complete with representative case studies, this authoritative resource provides a timely snapshot of the wide range of remote sensing applications in coastal issues to enhance the understanding of how increasing disturbances to our coastal regions are affecting the ecological dynamics, biological diversity, and ecosystem health of our coastal environments."--Publisher's description. Contents: Remote sensing of coastal environments: an overview / Yeqiao Wang -- Interferometric synthetic aperture radar (InSAR) study of coastal wetlands over southeastern Louisiana / Zhong Lu and Oh-Ig Kwoun -- Mangrove canopy 3D structure and ecosystem productivity using active remote sensing / Marc Simard, Lola E. Fatoyinbo, and Naiara Pinto -- Integration of LiDAR and historical maps to measure coastal change on a variety of time and spatial scales / Cheryl J. Hapke -- Coastal 3D change pattern analysis using LiDAR series data / Guoqing Zhou -- Mapping the onset and progression of marsh dieback / Elijah Ramsey III and Amina Rangoonwala -- Estimating chlorophyll condition in southern New England coastal waters from hyperspectral aircraft remote sensing / Darryl J. Keith -- Mapping salt mash vegetation by integrating hyperspectral and LiDAR remote sensing / Jiansheng Yang and Francisco Artigas -- Mapping salt marsh in Jamaica Bay and terrestrial vegetation in Fire Island National Seashore using QuickBird satellite data / Yeqiao Wang, Mark Christiano, and Michael Traber -- Object-based data integration and classification for high-resolution coastal mapping / Jie Shan and Ejaz Hussain -- True-color digital orthophotography data for mapping coastal impervious surface areas / Yuyu Zhou and Yeqiao Wang -- FORMOSAT-2 images in mapping of South Asia tsunami disaster / Ming-Der Yang, Tung-Ching Su, and An-Ming Wu -- Remote sensing and in situ measurements for delineation and assessment of coastal marshes and their constituent species / Martha S. Gilmore ... [et al.] -- Quantifying biophysical conditions of herbaceous wetland vegetation in Poyang Lake of coastal China via multitemporal SAR imagery and in situ measurements / Limin Yang ... [et al.] -- EO-1 advanced land imager data in submerged aquatic vegetation mapping / Eric Akins, Yeqiao Wang, and Yuyu Zhou -- Remote sensing applications used to inventory and monitor natural resources in North Atlantic coastal National Parks / Sara Stevens and Courtney Schupp -- Coastal area land-cover change analysis for Connecticut / James D. Hurd ... [et al.] -- Effects of increasing urban impervious surface on hydrology of coastal Rhode Island watersheds / Yuyu Zhou ... [et al.] -- Contemporary land-use/land-cover change in coastal Pearl River Delta and its impact on regional climate / Limin Yang ... [et al.] -- Geospatial information for sustainable development: a case study in coastal East Africa / Yeqiao Wang ... [et al.] Subjects: Coastal ecology--Remote sensing. Coasts--Remote sensing. Coastal zone management--Remote sensing. Environmental monitoring--Remote sensing. Notes: Includes bibliographical references and index. Series: Taylor & Francis series in remote sensing applications

Resilient cities: cities and adaptation to climate change - proceedings of the Global Forum 2010 / Konrad Otto-Zimmermann (ed.). LCCN: 2011929227 Meeting name: Resilient Cities Congress (2010: Bonn, Germany) Published/Created: Dordrecht: Springer, 2011. Description: xliv, 573 p.: ill.; 24 cm. ISBN: 9789400707849 (hbk.: alk. paper) 9400707843 (hbk.: alk. paper) 9789400707856 (e-book) 9400707851 (e-book) LC classification: QC981.8.G56 R45 2010 Contents: Part I. Introduction. Building the global adaptation community / Konrad Otto-Zimmerman -- Part II. Describing impacts at the local level. Introduction: describing impacts at the local level / Lucinda Fairhurst -- Understanding potential climate change impacts and adaptation options in India megacities / Diana Reckien, Martin Wildenberg, and Kaushik Deb -- Assessment of climate change-induced vulnerability to floods in Hyderabad, India, using remote sensing data / Oleksandr Kit, Matthias Lüdeke, and Diana Reckien -- The role of community resilience in adaptation to climate change: the urban poor in Jakarta, Indonesia / Mario Wilhelm -- Ecohealth and climate change: adaptation to flooding events in riverside secondary cities, West Africa / Guéladio Cissé, Brama Koné, Hampaté Bâ, Ibrahima Mbaye, Koffi Koba, Jürg Utzinger, and Marcel Tanner -- Climate change and adaptive capacity of women to water stress in urban centers of Nigeria: emerging concerns and reactions / Isaac B. Oluwatayo -- Urban climate strategies against future heat stress conditions / Lutz Katzschner -- Climate-proof urban transport planning: opportunities and challenges in developing cities / Urda Eichhorst, Daniel Bongardt, and Montserrat Miramontes -- Part III. Frameworks for local action, challenges and proactive recommendations. Introduction: frameworks for local action, challenges and proactive recommendations / Olivia Tusinski and Alice Balbo -- Inclusive adaptation: linking participatory learning and knowledge management to urban resilience / Michaela Hordijk and Isa Baud -- Urban adaptation planning and governance: challenges to emerging wisdom / JoAnn Carmin, David Dodman, Linda Harvey, Shaib Lwasa, and Patricia Romero-Lankao -- Urban climate change adaptation in the context of transformation: lessons from Vietnam / Matthias Garschagen and Frauke Krass -- Getting ready for a changing climate: supporting councillor's leadership role in adaptation / Joyce Wei Ka Lee -- Environmental learning from the ivory tower to the town square: the case of Trondheim, Norway / Annemie Wyckmans -- Plan B:altic: a social-ecological approach to climate change adaptation / Sonja Deppisch, Sanin Hasibović, and Meike Albers -- Governance recommendations for adaptation in European urban regions: results from five case studies and a European expert survey / Torsten Grothmann -- Creating resilient cities through empowered, deliberative participation / Janette Hartz-Karp and Hans-Peter Meister -- Part IV. Local strategies and actions in response to climate change. Introduction: Local strategies and actions in response to climate change / Monika Zimmermann and nathan Brettschneider -- Comparative research on the adaptation strategies of ten urban climate plans / Luc Vrolijks, Ashley Spatafore, and Anisha S. Mittal -- Managing regional climate mitigation and adaptation co-benefits and co-costs / Matthias Ruth -- Interdisciplinary and multi-institutional approaches to climate change adaptation / Virginie Moffet, Matthieu Alibert, and Caroline Larrivée -- Climate adaptation strategies: evidence from Latin American city-regions / Dirk Heinrichs and Kerstin Krellenberg -- Adaptation in UK cities: heading in the right direction? / Magnus Benzie, Alex Harvey, and Karen Miller -- Finding the balance: challenges and opportunities for climate change adaptation at different levels of English local government / Nadia Manasfi and Elizabeth Greenhalgh -- Local strategies for climate change adaptation: urban planning criteria for municipalities of the Basque country, Spain / Marta Olazabal, Efren Feliú, Borja Izaola, David Pon, Maria Pooley, Marimar Alonso-Martin, and Carlos Castillo -- Climate change adaptation in Dutch municipalities: risk perception and institutional capacity / Maya M. van den Berg -- Toward a more flood resilient urban environment: the Dutch multi-level safety approach to flood risk management / Berry Gersonius, William Veerbeek, Abdus Subhan, Karin Stone, and Chris Zevenbergen -- Climate change and city development: resilience strategies in Bonn / Rüdiger Wagner -- Integrated climate adaptation in Dresden: insights from flood prevention / Christian Korndörfer, Peter Teichmann, and Frank Frenzel -- Integrated climate protection program for the city of Chemnitz: climate diagnosis, climate change prognosis, and adaptation measures in the urban area/ Petra Schneider, Nicole Gottschalk, Ralf Löser, and Thomas Scharbrodt -- Adaptation strategies for the city of Cape Town: finding the balance within social-ecological complexity / Darryl Colenbrander, Gregg oelofse, Anton Cartwright, Howard Gold, and Sakhile Tsotsobe -- Sharing the Lagos megacity experience in the integrated management of sea level rise and flooding / Nicholas Ozor, Oguguah M. Ngozi, and Maximus Ugwuoke -- Community-based vulnerability assessment: Semarang, Indonesia / John Taylor -- Research news for climate compliant cities: the case of Ho Chi Minh City, Vietnam / Frank Schwarze, Andreas Gravert, Ronald Eckert, Ulrike Schinkel, and Ralf Kersten -- Building resilience to climate change through adaptive land use planning in Ho Chi Minh City, Vietnam / Harry Storch, Nigel Downes, Lutz Katzschner, and Nguyen Xuan Thinh -- Climate change risk management for a suburban local government: the case of Kogarah, Australia / Elisa Idris and Hartmut Fünfgeld -- Climate change as a survival strategy: soft infrastructure for urban resilience and adaptive capacity in Australia's coastal zones / Nan Chen and Peter Graham -- Curitiba, Brazil: a model for resilience in Latin America? / Peter Spathelf -- Part V. Approaches to climate change adaptation in cities. Introduction: Approaches to climate change adaptation in cities / Hartmut Fünfgeld -- Building resilience to climate change in Asian cities / Sarah Opitz-Stapleton, Laura Seraydarian, Karen MacClune, Greg Guibert, Sarah Reed, Fern Uennatornwaranggoon and Cristina Rumbaitis del Rio -- Local vulnerability assessment of climate change and its implications: the case of Gyeonggi-Do, Korea / Jaekyung Koh -- Ecosystem-based adaptation in the urban environment / Gabriel Grimsditch -- Urban agriculture and climate change adaptation: ensuring food security through adaptation / Marielle Dubbeling and Henk de Zeeuw -- Cooling the public realm: climate-resilient urban design / jeffrey Raven -- The role of transport in mitigation and adaptation to climate change impacts in urban areas / Giuseppe Inturri and Matteo Ignaccolo -- Urban green spaces: potentials and constraints for urban adaptation to climate change / Juliane Mathey, Stefanie Rössler, Iris Lehmann, and Anne Bräuer -- Urban climate framework: a system approach towards climate proof cities / Sonja Döpp, Fransje Hooimeijer, and Nienke Maas -- Part VI. Reality check: adaptation on the ground. Introduction: Reality check: adaptation on the ground / Olivia Tusinski -- Preparing for climate change while advancing local sustainability: a closer look at Miami-Dade County, Florida, USA / Natacha Seijas, Susanne M. Torriente, Nichole L. Hefty, and Melissa Stults -- Quito's climate change strategy: a response to climate change in the metropolitan district of Quito, Ecuador / Carolina Zambrano-Barragán, Othon Zevallos, Marcos Villacís, and Diego Enríquez -- Climate change implications for Dhaka City: a need for immediate measures to reduce vulnerability / Golam Rabbani, A. Atiq Rahman, and Nazria Islam -- Adaptation in practise: Durban, South Africa / Andrew Mather, Debra Roberts, and Geoffrey Tooley. Subjects: Global warming--Congresses. Climatic changes--Congresses. Climate and civilization--Congresses. Urban ecology (Biology)--Congresses. Urban ecology (Sociology)--Congresses. City planning--Congresses. Notes: Includes bibliographical references and index. Series: Local sustainability, 2211-2782; v. 1

Responding to climate change in New York State: the ClimAID integrated assessment for effective climate change adaptation in New York State: final report / editors, Cynthia Rosenzweig ... [et al.]. LCCN: 2012360737 Published/Created: Boston, Mass.: Published by Blackwell Pub. on behalf of the New York Academy of Sciences, 2011. Description: 1 v. (various pagings): col. ill., col. maps; 26 cm. Links: Publisher description http://www.loc.gov/catdir/enhancements/fy1208/2012360737-d.html

ISBN: 9781573318761 (pbk.) 1573318760 (pbk.) LC classification: QC903.2.U6 R47 2011 Q11 .N533b v. 1244 Portion of title: ClimAID integrated assessment for effective climate change adaptation in New York State Integrated assessment for effective climate change adaptation in New York State Related names: Rosenzweig, Cynthia. New York State Energy Research and Development Authority. Abstract: Climate change is already beginning to affect New York State, and these impacts are projected to grow. At the same time, the state has the ability to develop adaptation strategies to prepare for and respond to climate risks now and in the future. The ClimAID assessment provides information on climate change impacts and adaptation for eight sectors in New York State: water resources, coastal zones, ecosystems, agriculture, energy, transportation, telecommunications, and public health. Observed climate trends and future climate projections were developed for seven regions across the state. Within each of the sectors, climate risks, vulnerabilities, and adaptation strategies are identified. Integrating themes across all of the sectors are equity and environmental justice and economics. Case studies are used to examine specific vulnerabilities and potential adaptation strategies in each of the eight sectors. These case studies also illustrate the linkages among climate vulnerabilities, risks, and adaptation, and demonstrate specific monitoring needs. Stakeholder participation was critical to the ClimAID assessment process to ensure relevance to decision makers across the state-- Source other than Library of Congress.. Subjects: Climatic changes--New York (State) Climatic changes--Economic aspects--New York (State) Climate change mitigation--New York (State) Climatic changes--Government policy--New York (State) Climatic changes--Risk assessment--New York (State) Climatic changes--Risk management--New York (State) Notes: "Prepared for the New York State Energy Research and Development Authority." "This report was prepared by Columbia University, the City University of New York, and Cornell University in the course of performing work contracted for and sponsored by the New York State Energy Research and Development Authority."--P. iii. "NYSERDA 10851." Includes bibliographical references. Series: Annals of the New York Academy of Sciences, 0077-8923; v. 1244 NYSERDA (1995); 11-18.

Risk assessment of storms in coastal zones: case studies from Cartagena (Colombia) and Cadiz (Spain). LCCN: 2015932077 Published/Produced: New York, NY: Springer Berlin Heidelberg, 2015. Links: Table of contents only http://www.loc.gov/catdir/enhancements/fy1508/2015932077-t.html Publisher description http://www.loc.gov/catdir/enhancements/fy1508/2015932077-d.html ISBN: 9783319158433 Library of Congress Holdings Information not available.

Sedimentary coastal zones from high to low latitudes: similarities and differences / edited by I.P. Martini, University of Guelph, Canada and H.R. Wanless, University of Miami, USA. LCCN: 2015304444 Published/Produced: London: Published by the Geological Society, 2014. Description: vi, 600 pages; 26 cm. ISBN: 9781862393745 LC classification: GC380.15 .S395 2014 Related names: Martini, I. P. (Ireneo Peter), 1935- editor. Wanless, Harold R. (Harold Rogers), 1942- editor. Subjects: Marine sediments. Coast changes. Sediments (Geology) Notes: Includes bibliographical references and index. Series: Geological Society special publication, 0305-8719; no. 388

Tropical deltas and coastal zones: food production, communities and environment at the land-water interface / edited by Chu Thai Hoanh ... [et al.]. LCCN: 2009046423 Meeting name: Delta 2007 (2007: Ban Bang Saen, Thailand) Published/Created: Wallingford, Oxfordshire, UK; Cambridge, MA: CABI, c2010. Description: xvii, 477 p.: ill., maps; 26 cm. ISBN: 9781845936181 (hc: alk. paper) 1845936183 LC classification: SH134.6 .D45 2007 Related names: Chu, Thai Hoanh, 1949- International Water Management Institute. WorldFish Center. International Rice Research Institute. FAO Regional Office for Asia and the Pacific. Challenge Program on Water and Food. Subjects: Aquaculture--Tropics--Congresses. Fisheries--Tropics--Congresses. Water use--Developing countries--Management--Congresses. Deltas--Developing countries--Congresses. Deltas--Tropics--Congresses. Coastal zone management--Developing countries--Congresses. Notes: "In association with the International Water Management Institute (IWMI), the WorldFish Center, the International Rice Research Institute (IRRI), Food and Agriculture Organization of the United Nations (FAO)--Regional Office for Asia and the Pacific, and CGIAR Challenge Program on Water and Food (CPWF)." Includes bibliographical references and index. Series: Comprehensive assessment of water management in agriculture series; v. 9

INDEX

M

N

O

P

Q

R

S

T

U

V

W